Convex Surfaces

HERBERT BUSEMANN

DOVER PUBLICATIONS, INC.
Mineola, New York

Bibliographical Note

This Dover edition, first published in 2008, is an unabridged republication of the work originally published in 1958 by Interscience Publishers, Inc., New York.

Library of Congress Cataloging-in-Publication Data

Busemann, Herbert, 1905–
Convex surfaces / Herbert Busemann.
p. cm.
Originally published: New York : Interscience Publishers, 1958, in series: Interscience tracts in pure and applied mathematics ; no. 6.
ISBN-13: 978-0-486-46243-1
ISBN-10: 0-486-46243-9
1. Convex surfaces. I. Title.

QA649.B78 2007
516.3'62—dc22

2007010608

Manufactured in the United States of America
Dover Publications, Inc., 31 East 2nd Street, Mineola, N.Y. 11501

PREFACE

The purpose of this tract is to acquaint the mathematical public with a subject, convex surfaces, which during the past 25 years has experienced a striking and beautiful development, principally in Russia, but has remained largely unknown, at least in the U.S.A.

The book grew out of lectures which the author gave, upon the request of his colleagues, at the Summer Institute for Differential Geometry in the Large at Seattle in 1956. These lectures were, of necessity, hastily prepared and this is, so to speak, the version which the author would have liked to present. It is not an encyclopedic account with complete proofs, which would have required something like 800 pages, and would have frightened a potential friend away from, rather than have acquainted him with, the theory of convex surfaces. Consequently, some indications about the method of presentation will be welcome.

The reader will find a self-contained description of the main results of the theory: all definitions are given and all theorems are formulated precisely. The number of complete proofs in the first two chapters is greater than in the last two, for the following reason: Chapters I and II form, in many respects, a natural complement to Bonnesen and Fenchel's *Theorie der konvexen Körper* (1934). They deal largely with subjects conjectured or suggested in that report and are written in this spirit: proofs found in the report are not repeated. This material appears here for the first time in book form.

Most of the content of the last two chapters can be found in books, mainly in A. D. Alexandrov's *Die innere Geometrie der konvexen Flächen.* The major part of the present book, also of Chapters I and II, is in fact due to Alexandrov, and a great deal of the rest was stimulated by him. Many theorems, in particular of Chapter III, deal with general forms of well-known classical theorems and the absence of proofs, which are usually quite long, will not be found disturbing.

The prerequisites vary in different parts of the tract. A thorough familiarity with the classical differential geometry of surfaces is assumed throughout, and very little else is necessary for Chapter III and large parts of Chapter IV. The theory of real variables, in particular of set functions, is used in Chapters I and II. Chapter II, which is independent of the rest, uses Minkowski's mixed volumes (but the definitions and theorems, drawn from this theory, are stated), and also requires some knowledge of eigenvalues of quadratic forms and of integral equations. The last part of Section 5 and Section 23 use the theory of elliptic partial differential equations.

The author dislikes books which jump into an exposition without informing the reader beforehand of the content. In the present case, familiarity with the classical theory makes most of the section headings sufficiently descriptive, and where this is not the case, the first few lines of a section will inform the reader of its purpose. Only the headings of Sections 21 and 22 require some explanation. There has been a conspicuous lack, in all languages, of a term which briefly describes the following situation: a certain two-dimensional metric is given either abstractly or as the intrinsic metric of a convex surface, and there is exactly one realization of this metric within a given class C of convex surfaces. In that case we call here the given metric monotypic in C.

It would have been futile to attempt compiling a complete bibliography, which would doubtless have been useful in some respects. Most of the work was done in Russia, at times in quite inaccessible journals, and occasional references show that not all papers have reached the West.

Since completeness was out of question, the Literature, on pp. 189–194, was kept strictly in line with the purpose of this tract: apart from a few historically particularly significant papers, it contains nothing but the most convenient sources that the author could provide. For example, many results appeared first as brief, usually Doklady, notes, then in papers and finally in books, and often only the latter is quoted. Thus the date of a quotation frequently bears little relation to the date of the discovery. This was feasible because there do not seem to be any priority disputes in the field.

The author acknowledges his gratitude to the National Science Foundation: it subsidized not only the aforementioned Summer Institute but also the later work on this tract.

For assistance in proof reading, the author is indebted to Mr. J. W. Bergquist.

HERBERT BUSEMANN

UNIVERSITY OF SOUTHERN CALIFORNIA, LOS ANGELES

After this book was completely printed, N. W. Efimow's *Flachenverbiegung im Grossen*, Berlin, 1957, reached the author. The first half of that book is a translation of Efimow's *Qualitative Questions in the Theory of Deformation of Surfaces* (Russian), Uspehi Matem. Nauk N.S., vol. 3 (1948), pp. 47–158; the second half is a report on the later developments by E. Rembs and K. P. Grotemeyer. The book overlaps our Chapter IV. The following parts are useful as additions to the present book: an outline of Alexandrov's first proof for the realization of polyhedral metrics (Sections 6–9); a proof of the same theorem by Lyusternik using the Weyl-Lewy result on the realization of analytic metrics (Section 12). A more detailed outline (Section 68), than is given here, of Pogorelov's general monotypy theorem may interest those who do not care to read the complete proof in Pogorelov [7] (German). Finally, at various places simpler proofs are found which become available for results discussed here, when smoothness hypotheses are introduced.

The author has just learned that a translation of the frequently-quoted book Pogorelov [3] has appeared as: A. W. Pogorelow, *Die Verbiegung konvexer Flächen*, Akademie-Verlag, Berlin, 1958, 135 pp.

CONTENTS

CHAPTER I

Extrinsic Geometry

CHAPTER II

The Brunn-Minkowski Theory and Its Applications

CHAPTER III

Intrinsic Geometry

CHAPTER IV

Realization of Intrinsic Metrics

CHAPTER V

Conclusion

CHAPTER I

Extrinsic Geometry

1. Notations, terminology, basic facts

The terminology used in the literature on convex bodies or surfaces is not quite uniform. An agreement, at the outset, in this respect as well as on the notations for certain simple recurring concepts will therefore prove useful.

E^n is the n-dimensional euclidean space. An *r-flat*, $0 \leqq r \leqq n$, in E^n is an r-dimensional linear subvariety of E^n; the 0-, 1- and $(n-1)$-flats are, of course, also called points, lines, and hyperplanes (planes for $n = 3$).

In a finite dimensional space like E^n no actual disagreement on the topology for its r-flats can exist. However, to satisfy readers troubled by the difficulties or ambiguities occurring in infinite-dimensional spaces, we mention that convergence of point sets may in this book always be interpreted in the sense of *Hausdorff's closed limit.* [1]

If Cartesian coordinates $x_1, \ldots, x_n$ have been introduced in E^n then a means the point with coordinates $a_1, \ldots, a_n$ and $|a - b| = [\sum(a_i - b_i)^2]^{1/2}$ is the *distance* of a and b,

$$U(p, \varrho) \text{ is the open sphere } |p - x| < \varrho,$$
$$S(p, \varrho) \text{ is its boundary } |p - x| = \varrho.$$

The closure $\overline{U}(p, \varrho) = \overline{U(p, \varrho)} = U(p, \varrho) \cup S(p, \varrho)$ of $U(p, \varrho)$ is the solid sphere or *ball* $|p - x| \leqq \varrho$.

Finally, if a and b are two distinct points then the points of the

[1] See for example Busemann [3], Section 3, where it is shown that the closed sets in any finitely compact metric space form, with an appropriate distance, again a finitely compact space in which convergence is equivalent to convergence of sets in the original space in terms of Hausdorff's closed limit.

form $(1 - \varrho)\, a + \varrho\, b$ form for
$-\infty < \varrho < \infty$ the *line* $L(a, b)$ through a and b,
$0 \leqq \varrho < \infty$ the *ray* $R(a, b)$ from a through b,
$0 \leqq \varrho \leqq 1$ the (*closed*) *segment* $E(a, b)$ from a to b,
$0 \leqq \varrho < 1$ the *half-open segment* $E(a, b\}$ from a to b.

With $L^{+}(a, b)$, and correspondingly for the other sets, we mean that *orientation* of $L(a, b)$ for which increasing ϱ yields a traversal in the positive sense.

A non-empty set K in E^n is *convex* if it contains with any two points x, y the entire segment $E(x, y)$. Since the intersection of any aggregate of convex sets in E^n is empty or convex and the r-flats are convex, there is a smallest r-flat, say V_m of dimension m, which contains K. Then K contains $m + 1$ points which do not lie in an $(m - 1)$-flat, and hence the simplex spanned by these points. Therefore K is a convex set with interior points relative to V_m and K has *dimension* m, either by definition or in the sense of dimension theory.

The shapes of all convex sets in the euclidean spaces will therefore be known if we know the possible shapes of an n-dimensional convex set K in E^n. Let p be an interior point of K and $U(p, \varrho) \subset K$. If x is an arbitrary point of K then $E(z, x) \subset K$ for any $z \in U(p, \varrho)$. Since $W = \bigcup_{z \in U(q, \rho)} E(z, x\}$ is open, every point of $E(p, x\}$ is an interior point of K. Therefore:

(1.1) *The interior points of K form a convex set.*

(1.2) *The closure $\bar{K}$ of K is convex.*

It also follows that for every $y \in S(p, \varrho)$ the ray $R(p, y)$ either lies entirely in K or intersects the boundary $B(K)$ of K in exactly one point $b(y)$. In the latter case all points of $R^{+}(p, b(y))$ following $b(y)$ lie outside K and all points of $E(p, b(y)\}$ are interior points of K.

We put $\delta(y) = |\, p - b(y)\,|$ and $\delta(y) = \infty$ if $R(p, y) \subset K$. It follows easily from $W \subset K$ for $x \in K$ that $\delta(y)$ is a continuous function of y. Hence

(1.3) *The interior of an n-dimensional convex set K is homeomorphic to E^n.*

If K is bounded then its boundary $B(K)$ is homeomorphic to the $(n-1)$-dimensional sphere S^{n-1}. When K is not bounded we consider first the case where K does not contain a complete line. The set M of those y on $S(p, \varrho)$ for which $\delta(y) = \infty$ is then non-empty and closed, and contains no two antipodal points of $S(p, \varrho)$. The union of all rays $R(p, y) \subset K$ is evidently a convex, possibly degenerate, cone with apex p, and the intersection of this cone with $\overline{U}(p, \varrho)$ is a bounded closed convex set. Therefore $S(p, \varrho) - M$ and hence $B(K)$ is homeomorphic to E^{n-1}.

If K contains a complete line then it contains a flat V_r of maximal dimension r. If $r=n$ then $K=E^n$. Assume $1 \leqq r \leqq n-1$ and take any $(n-r)$-flat V_{n-r} normal to V_r. Then $K' = V_{n-r} \cap K$ is an $(n-r)$-dimensional convex set in V_{n-r} which does not contain a complete line, so that its boundary $B(K')$ is by the preceding discussion homeomorphic to S^{n-r-1} or to E^{n-r-1}. Since K contains for $q \in K'$ and $x \in V_r$ the segment $E(q, x)$, the closure $\overline{K}$ of K contains with any $q \in K'$ the r-flat through q parallel to V_r. Therefore $B(K)$ consists of the r-flats parallel to V_r through points of $B(K')$ and is therefore homeomorphic either to the product $E^{n-r-1} \times E^r$, i.e., to E^{n-1} or to $S^{n-r-1} \times E^r$, which may be regarded as a cylinder with r-dimensional generators.

(1.4) *If the boundary $B(K)$ of an n-dimensional convex set K in E^n is not empty ($K \neq E^n$) then it is homeomorphic either to E^{n-1} or to S^{n-1} or to the product $S^{n-r-1} \times E^r$, $1 \leqq r \leqq n-1$.*

The product $S^{n-r-1} \times E^r$ is connected except for $r = n-1$, where $B(K)$ consists of two parallel hyperplanes.

A *(complete) convex hypersurface* in E^n is the boundary of an n-dimensional convex set K in E^n provided it is non-empty and connected, i.e., $B(K)$ is a convex hypersurface unless K is the whole space or a set bounded by two parallel hyperplanes. Since convex surfaces are the primary subject of this book we reformulate (1.4):

(1.5) *A (complete) convex hypersurface in E^n is either homeomorphic to S^{n-1} or to E^{n-1}, or to a product $S^{n-r-1} \times E^r$, $1 \leqq r \leqq n-2$ and is, respectively, called closed or open or cylindrical.*

For $n = 2, 3$ we use, of course, the terms curve and surface instead of hypersurface. A (complete) convex curve is homeomorphic to a line or to a circle, and a (complete) convex surface is homeomorphic to a plane, a two-sphere or to an ordinary cylinder. The term *convex hypersurface* will also be used for *relative open connected subsets of complete convex hypersurfaces.* The parentheses indicate that "complete" will be omitted when the distinction is inessential or clear from the context.

A *supporting r-flat* of the n-dimensional convex set K in E^n is an r-flat which contains points of $\bar{K}$ but no interior points of K. The supporting 0-flats are the points of $B(K)$, and the supporting $(n-1)$-flats, simply called *supporting planes*, are the hyperplanes which contain points of K but do not separate any two points of K. This property is used as *definition for supporting planes of a convex set without interior points*, but for obvious reasons supporting r-flats with $r < n - 1$ are not defined for such sets. Notice the following obvious, but useful, fact:

(1.6) *If $V_1, V_2, \ldots$ are supporting r-flats of K and tend to an r-flat V which contains points of $\bar{K}$* (when K is bounded $\lim V_n$ automatically contains points of $\bar{K}$), *then V is a supporting r-flat of K.*

Next we show

(1.7) *If K_1 is a bounded convex set and K_2 any convex set and $\bar{K}_1 \cap \bar{K}_2 = 0$, then points $f_i \in \bar{K}_i$ exist such that*

$$| f_1 - f_2 | = \min_{x_i \in \bar{K}_i} | x_1 - x_2 |$$

If P_1, P_2, P are the hyperplanes normal to $L(f_1, f_2)$ at f_1, f_2 and $(f_1 + f_2)/2$ then P_i is a supporting plane of K_i and P separates K_1 from K_2. If K_1 is a point f_1 then f_2 is unique.

The existence of f_1 and f_2 follows from the boundedness of K_1. If P_2 were not a supporting plane of K_2 then there would be a point q of K_2 on the same side of P_2 as f_1. But the angle $\angle q f_2 f_1$ is acute, hence $| f_1 - (1 - \varrho) f_2 - \varrho q | < | f_2 - f_1 |$ for small positive ϱ and $(1 - \varrho) f_2 + \varrho q \in K_2$. It is clear that P separates K_1 from K_2 and has, in fact, distance $| f_1 - f_2 |/2$ from K_i.

If $K_1 = f_1$ then f_2 is called a *foot* of f_1 on K_2. A second foot g of f_1 on K_2 would lead to the contradiction $(f_2 + g)/2 \in K_2$ and $|f_1 - (f_2 + g)/2| < |f_1 - f_2| = |f_1 - g|$.

We use (1.7) to prove:

(1.8) *If an arbitrary convex set K' and an n-dimensional convex set K in E^n have a common boundary point p and K' contains no interior points of K, then K and K' possess at p a common supporting plane. Consequently, a supporting r-flat of K, in particular a point of $B(K)$, lies in a supporting plane of K.*

Choose any interior point q of K and call K_n the set obtained from $\bar{K}$ by the dilation with center q and factor $1 - n^{-1}$. If $K'_n = \bar{K}' \cap \bar{U}(p, n)$, then $K'_n \cap K_n = 0$ and (1.7) furnishes a hyperplane P_n which separates K'_n from K_n and contains, therefore, a point of the segment $E(p, p_n)$ from p to the image p_n of p under the dilation. Since $|p - p_n| = |p - q|/n$, there is a subsequence $\{P_\nu\}$ of $\{P_n\}$ which tends to a hyperplane $P \supset p$. Then P cannot separate two points a_1, a_2 of K because there are points $a_i^\nu \in K_\nu$ which tend to a_i and P_ν would separate a_1^ν from a_2^ν for large ν. Similarly, P does not separate two points a'_1 and a'_2 of K', because $a'_i \in K'_n$ for large n, and P_ν does not separate a'_1 from a'_2 for $\nu > n$.

A *supporting half space* of the convex set K is a closed half space bounded by a supporting plane of K and containing K. A corollary of (1.8) is:

(1.9) *If $K \neq E^n$ then $\bar{K}$ is the intersection of the supporting half spaces of K.*

Observe the further corollary of (1.8):

(1.10) If a_ν and b_ν, $\nu = 1, 2, \ldots$, are two distinct points on the boundary $B(K)$ of the n-dimensional set K which tend to the same point p, then the limit L of any converging subsequence of $L(a_\nu, b_\nu)$ lies in a supporting plane of K at p.

For L is a supporting line of K because all interior points of K on $L(a_\nu, b_\nu)$ must lie on $E(a_\nu, b_\nu)$.

Let C be a convex hypersurface bounding the convex set K, which is uniquely determined by C except in the trivial case where C is a hyperplane. The interior of K and the supporting planes of K are also called the interior and the supporting planes of C.

The intersection of all supporting half spaces of K bounded by supporting planes of K at a fixed point b of C is a convex cone T_0 with interior points whose boundary T is, by definition, *the tangent cone of K or C at b.* If T is a hyperplane, then it is the unique supporting plane of C at b. We then also say that T is the *tangent plane* of C at b and that *C is differentiable* at b. We conclude from (1.6):

(1.11) *If C is differentiable at the points $x \in M \subset C$ then the tangent plane of C at x depends continuously on x.*

A convex hypersurface C is locally representable in the form $z = f(x_1, \ldots, x_{n-1})$. This is quite obvious if we are satisfied with oblique coordinates. It suffices to choose an arbitrary supporting plane of C at a given point b as $z = 0$ and any line L through b and an interior point of C as z-axis. Then (1.10) shows that a line parallel and close to L intersects C exactly once in a neighborhood of b. To see that rectangular coordinates with this property exist, we form the complementary or polar cone T' to T which consists of the rays normal at b to the supporting plane of T and in the exterior of T. (If T is a plane then T' consists of the ray normal to T and in the exterior of K.) It is easy to see that T' is convex, see K. p. 4. Because $T_0 \supset K$ the cone T' possesses at b a supporting plane P touching T' only at b. The line L normal to P contains interior points of K, with P as plane $z = 0$ and L pointing into the interior of K as positive z-axis we have a rectangular coordinate system which satisfies the following theorem:

(1.12) THEOREM. *For a given point b on a convex hypersurface C there is a neighborhood C_b of b on C and a system of rectangular coordinates $x_1, x_2, \ldots, x_{n-1}$, z with b as origin such that C_b is representable in the form*

$$z = f(x_1, \ldots, x_{n-1}) = f(x) \geqq 0, \; f(0) = 0, \; |x| \leqq \delta, \; \delta > 0.$$

Moreover $f(x)$ *is a convex function of* x, [2] *and its difference quotients* $|f(x) - f(y)|/|x - y|$, $x \neq y$, *are bounded.*

If (x_0, z_0) *is a point of* C_b *where* C (or C_0) *is differentiable then* $f(x)$ *possesses at* x_0 *a differential.*

The convexity of $f(x)$ follows at once from the convexity of the set $|x| \leqq \delta$, $z \geqq f(x)$ in E^n. The boundedness of the difference quotients is a consequence of (1.10), because the z-axis contains interior points of K, and so is the existence of the differential. For if $z = 0$ is the tangent plane of C at b, then (1.10) implies the much stronger property that for a given $\epsilon > 0$ and a suitable $\epsilon' > 0$ the inequality

$$|f(x) - f(y)|/|x - y| < \epsilon \text{ holds for } |x| < \epsilon' \text{ and } |y| < \epsilon'.$$

The boundedness of the difference quotients shows that we need not worry regarding the concept of area of a convex hypersurface. Also, a set of measure 0 on $\bar{C}_b$ has a projection of measure 0 in $|x| \leqq \delta$.

2. Convex curves

In this and the next section we discuss the differentiability properties of convex hypersurfaces in greater detail. Since these are local we may consider partial surfaces representable in the form (1.12).

For a function $f(x)$ of one variable x, $a \leqq x \leqq b$, we use

$$f'_r(x) = \lim_{h \to 0+} \frac{f(x+h) - f(x)}{h}, \quad f'_l(x) = \lim_{h \to 0-} \frac{f(x+h) - f(x)}{h}$$

as symbols for the right and left derivatives.

If $y = f(x)$ is a convex curve, then for $x_1 < x_2$, $x'_1 < x'_2$, $x_1 \leqq x'_1$, and $x_2 \leqq x'_2$

$$\frac{f(x_1) - f(x_2)}{x_1 - x_2} \leqq \frac{f(x'_1) - f(x'_2)}{x'_1 - x'_2},$$

[2] The function $f(x)$ defined in a convex set M of the $(x_1, \ldots, x_{n-1})$-space is convex if $x, y \in M$ implies $f((1-\varrho)x + \varrho y) \leqq (1-\varrho)f(x) + \varrho f(y)$ for $0 < \varrho < 1$, see K., p. 18.

hence $f'_r(x)$ and $f'_l(x)$ exist; and if $f'_m(x)$ is the slope of an arbitrary supporting line at x, then

(2.1) $f'_l(x) \leqq f'_m(x) \leqq f'_r(x)$ and $f'_m(x_1) \leqq f'_m(x_2)$ for $x_1 < x_2$.

The monotoneity of $f'_m(x)$ implies that the *number of points where $f'(x)$ does not exist is, at most, denumerable.* It also implies that $f'_m(x)$ *has almost everywhere a derivative,* and that all points where the derivatives of $f(x)$ and $f'_m(x)$ exist, the latter has the same value for different choices of $f'_m(x)$. We therefore simply denote it as the (finite) second derivative $f''(x)$ of $f(x)$ at x. We remember for later purposes that:

$$(2.2) \qquad f''(x) = \lim_{h\to 0} [f(x+h) + f(x-h) - 2f(x)]h^{-2}$$

We now investigate the curvature of convex curves. The radius of the circle (possibly ∞) which passes through the points (x_0, y_0) and $(x_0 + h, y_0 + k)$ and has slope m at (x_0, y_0) equals

$$(2.3) \qquad r = \pm(1 + m^2)^{1/2}\left(1 + \frac{k^2}{h^2}\right)\frac{h}{2}\left(\frac{k}{h} - m\right)^{-1}$$

Its center lies, of course, on the line $(x - x_0) = -m(y - y_0)$. A line which passes through $(x_0 + h, y_0 + k)$ and is normal to the line with slope m_h intersects $x = -my$ at a point whose distance from (x_0, y_0) equals

$$(2.4) \qquad R = \pm(1 + m^2)^{1/2}\left(1 + \frac{k}{h}m_h\right)\left(\frac{m - m_h}{h}\right)^{-1}$$

Now consider a continuous curve $y = f(x)$ which has for $x = x_0$ a derivative. The radius $r_h(x_0)$ of the circle which has the same tangent as the curve at $(x_0, y_0 = f(x_0)) = p_0$ and passes through $p_h = (x_0 + h, f(x_0 + h) = y_0 + k)$, $h > 0$, has by (2.3) a right limit $\lim_{h\to 0+} r_h$ if the so-called second right de la Vallée-Poussin derivative

$$(2.5) \qquad D''_r f(x_0) = \lim_{h\to 0+} \frac{2}{h}\left(\frac{f(x_0 + h) - f(x_0)}{h} - f'(x_0)\right)$$

exists and

$$\lim_{h\to 0+} r_h(x_0) = \pm\,(1 + f'^2(x_0))^{3/2}/D''_r f(x_0). \tag{2.6}$$

If $f(x)$ possesses a derivative in a neighborhood of x_0 then the distance $R_h(x_0)$ from p_0 of the intersection of the normals to $y = f(x)$ at p_0 and p_h has a limit if $f'(x)$ possesses at x_0 the right derivative

$$f''_r(x_0) = \lim_{h\to 0+} h^{-1}(f'(x_0 + h) - f'(x_0)) \tag{2.7}$$

and

$$\lim_{h\to 0+} R_h(x_0) = \pm(1 + f'^2(x_0))^{3/2}/f''_r(x_0). \tag{2.8}$$

It follows from Taylor's theorem that *the existence of* $f''_r(x_0)$ *implies the existence of* $D''_r f(x_0)$ and that then the two derivatives are equal. The function $y = x^3 \sin x^{-1}$ shows that the converse need not hold; in fact $D'' f(x_0) = D''_r f(x_0) = D''_l f(x_0)$ may exist in a set of positive measure without $f''(x_0)$ existing anywhere, see Denjoy [1].

If, for any differentiable $f(x)$, the circle of radius $r_h(x_0)$ touching $y = f(x)$ at p_0 and passing through p_h has center a_h, then the arc $x_0 \leqq x \leqq x_0 + h$ of $y = f(x)$ contains a point $p_{h'}$, $0 < h' < h$ whose distance from a_h is maximal or minimal. The normal to the curve at this point passes through a_h so that $R_{h'}(x_0) = r_h(x_0)$. Hence, and similarly,

$$\begin{aligned} \limsup_{h\to 0+} R_h(x_0) &\geqq \limsup_{h\to 0+} r_h(x_0), \\ \liminf_{h\to 0+} R_h(x_0) &\leqq \liminf_{h\to 0+} r_h(x_0). \end{aligned} \tag{2.9}$$

If $f(x)$ is a convex function and $f'(x_0)$ exists, then we will interpret $f'(x_0 + h)$ or m_h in (2.4) as the slope $f'_m(x_0 + h)$ of any supporting line at p_h. Because $f'_m(x_0 + h)$ is monotone $\limsup_{h\to 0+} R_h(x_0)$ and $\liminf_{h\to 0+} R_h(x_0)$ are independent of the choice of $f'_m(x_0 + h)$. Then (2.9) holds and, if

$$f''_r(x_0) = \lim_{h\to 0+} h^{-1}(f'_m(x_0 + h) - f'(x_0)),$$

exists for one choice of $f'_m(x_0 + h)$, it exists for all and is independent of the choice.

Jessen [1] discovered the surprising fact that *for convex* $f(x)$ *the existence of* $D''_r f(x_0)$ *and that of* $f''_r(x_0)$ *are equivalent.* He deduced this from the following more general result:

(2.10) THEOREM. *Let the convex curve* $y = f(x)$ *possess a tangent at* $[x_0, f(x_0)]$ *and put*

$$r_s = \limsup_{h\to 0+} r_k(x_0),\ r_i = \liminf_{h\to 0+} r_h(x_0).$$

Then

$$r_s - r_s^{1/2}(r_s - r_i)^{1/2} \leqq \liminf_{h\to 0+} R_h \leqq \limsup_{h\to 0+} R_h \leqq r_s + r_s^{1/2}(r_s - r_i)^{1/2},$$

where for $r_s = \infty$ *the left member means* $r_i/2$.

For a proof we consider the two circles

$$A\colon x^2 + (y - \alpha)^2 = \alpha^2,\ B\colon x^2 + (y - \beta)^2 = \beta^2,\ \alpha > \beta > 0.$$

From $p_1 = (\alpha, \alpha) \in A$ we draw the tangent to B which has the greater slope and call p_2 its second intersection with A. We proceed with p_2 as with p_1, i.e., we draw the tangent different from $L(p_1, p_2)$ to B; it intersects A a second time at p_3. Continuing in this manner we obtain an infinite polygon $\cup E(p_i, p_{i+1})$ which together with the arc $-\alpha < x < 0$ of A forms a convex curve D of the form $y = f(x)$ which has at $p_0 = (0,0)$ the x-axis as tangent. Clearly for this curve

$$\alpha = \limsup_{h\to 0+} r_h(0),\ \beta = \liminf_{h\to 0+} r_h(0)$$

If $p_i = (h_i, k_i)$ and δ_i is the acute angle formed by $L(p_{i-1}, p_i)$ and the x-axis, then the minimum and maximum for $R_h(0)$ on the interval $h_i < h < h_{i-1}$ are reached at the end points, hence from (2.4)

$$\limsup_{h\to 0+} R_h(0) = \lim_{i\to\infty}\left(1 + \frac{k_i}{h_i}\tan\delta_{i+1}\right)\frac{h_i}{\tan\delta_{i+1}} = \lim\frac{|p_i|}{\delta_{i+1}}$$

$$\liminf_{h\to 0+} R_h(0) = \lim\left(1 + \frac{k_i}{h_i}\tan\delta_i\right)\frac{h_i}{\tan\delta_i} = \lim\frac{|p_i|}{\delta_i}$$

because $|p_i|\,h_i^{-1} = (1 + k_i^2\,h_i^{-2})^{1/2} \to 1.$

Let $q_i = (h'_i, k'_i)$ be the foot of the center $a = (0, \alpha)$ of A on $L(p_{i-1}, p_i)$. Then $|a - q_i| \to \alpha$, $h'_i = |a - q_i| \sin \delta_i$, hence

$$\alpha = \lim \frac{h'_i}{\sin \delta_i} = \lim \frac{|q_i|}{\delta_i}.$$

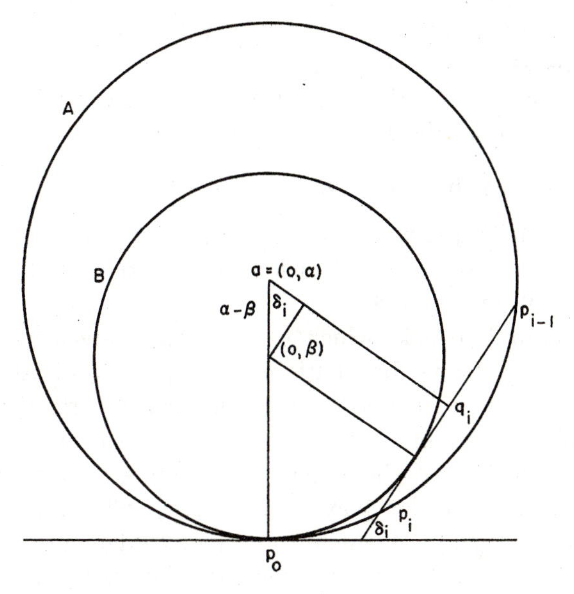

Figure 1.

Moreover, see Figure 1,

$$|q_i - p_{i-1}|^2 = |q_i - p_i|^2 = \alpha^2 - [\beta + (\alpha - \beta)\cos \delta_i]^2$$
$$= (\alpha - \beta)(1 - \cos \delta_i)\,[\alpha(1 + \cos \delta_i) + \beta(1 - \cos \delta_i)]$$

and

$$\lim \delta_i^{-1}\,|q_i - p_{i-1}| = \lim \delta_i^{-1}\,|q_i - p_i| = (\alpha - \beta)^{1/2}\,\alpha^{1/2}.$$

It follows from $< p_0 p_i p_{i-1} \to \pi$ that

$$\lim \delta_i^{-1}|p_{i-1}| = \lim \delta_i^{-1}|q_i| + \lim \delta_i^{-1}|q_i - p_{i-1}| = \alpha + \alpha^{1/2}(\alpha - \beta)^{1/2}$$
$$\lim \delta_i^{-1}|p_i| = \lim \delta_i^{-1}|q_i| - \lim \delta_i^{-1}|q_i - p_i| = \alpha - \alpha^{1/2}(\alpha - \beta)^{1/2}.$$

For the curve D we have therefore

$$\limsup_{h\to 0+} R_h(0) = r_s + r_s^{1/2}(r_s - r_i)^{1/2} = \alpha + \alpha^{1/2}(\alpha - \beta)^{1/2}.$$

$$\liminf_{h\to 0+} R_h(0) = r_s - r_s^{1/2}(r_s - r_i)^{1/2} = \alpha - \alpha^{1/2}(\alpha - \beta)^{1/2}.$$

Clearly, D yields for all convex curves between A and B the extreme values for the limits of $R_h(0)$.

If C is any convex curve in $y > 0$ with $0 < r_i \leqq r_s < \infty$ and the x-axis as tangent at p_0, let $0 < \beta < r_i$ and $\alpha > r_s$. Then a certain subarc of C to the right of p_0 will lie between A and B and we conclude that the limits of $R_h(0)$ for C satisfy

$$\alpha-\alpha^{1/2}(\alpha-\beta)^{1/2} \leqq \liminf R_h(0) \leqq \limsup R_h(0) \leqq \alpha + \alpha^{1/2}(\alpha-\beta)^{1/2}.$$

For $\beta \uparrow r_i$ and $\alpha \downarrow r_s$ we obtain the assertion. A rather obvious modification of this method yields the assertion for the limiting cases $r_i = 0$ and $r_s = \infty$.

It is natural to ask, which parts of the convexity condition produce these strong differentiability properties; in other words, we ask for a more general geometric property than convexity which has similar implications. The first idea which comes to mind is finiteness of the order. To include curves containing segments, in particular polygons, we say, following Hjelmslev, that a curve $(x(t), y(t))$ which is locally a Jordan curve has order $\leqq n$ if the intersection of any line with the curve has at most n components. A curve has order n if its order is $\leqq n$ but not $\leqq n - 1$.

According to Marchaud [1] *a curve of order* $\leqq 3$ is the union of at most four (in general not complete) convex curves, and *has therefore all the differentiability properties of convex curves. Surprisingly enough, curves of order* $\geqq 4$ *need not have these properties*, see Haupt [1] and B.F. [3].

However, finiteness of the class, and an even weaker condition prove adequate: Call oriented tangent at p of an oriented curve any oriented line T which is limit of lines $L^+(q, r)$ where q and r, respectively, precede and follow p on the curve and tend to p. For any φ, $0 \leqq \varphi < 2\pi$ let $N(\varphi)$ be the number of points on the curve where an oriented tangent exists which forms the angle φ

with a fixed oriented line. If $N(\varphi)$ is summable, i.e., if the total curvature

$$k = \int_0^{2\pi} N(\varphi) d\varphi$$

is finite, then the curve has all differentiability properties of convex curves (with the exception of (2.10) and its implication that the existence of $D''_r f(x)$ implies that of $f''_r(x)$). For details see B. F. [3].

We will see later that the situation is analogous for surfaces: the surfaces of bounded curvature, and not the surfaces of finite order, share the essential differential geometric properties with convex surfaces.

3. *The theorems of Meusnier and Euler*

The considerations of this section will be essentially local, so that in the proofs we may restrict ourselves to a neighborhood C_b of a given point b on a hypersurface C which satisfies (1.12).

(3.1) *A convex hypersurface is almost everywhere differentiable.*

This theorem is due to Reidemeister [1]. We know that it holds for curves, i.e., in E^2. Assuming that (3.1) holds in E^{n-1}, $n \geqq 3$, we will prove it in E^n.

With the notations of (1.12) the set $z = f(x)$, $x_i = \alpha$, $|\alpha| < \delta$, is a convex hypersurface H_α^i relative to $x_i = \alpha$, i.e., in an E^{n-1}. The set of those points in $x_i = \alpha$ and $z = 0$ corresponding to points where H_α^i is not differentiable, has by assumption $(n-2)$-dimensional measure 0. Therefore the — obviously measurable — set N_t of the $(\bar{x}, 0)$ where $H_{\bar{x}_1}^1$ or $H_{\bar{x}_2}^2$ is not differentiable, has $(n-1)$-dimensional measure 0. If $(\bar{x}, 0)$ is not in N_t then the tangent cone of C_b at $(\bar{x}, f(\bar{x}))$ contains the $(n-2)$-flats tangent to $H_{\bar{x}_1}^1$ and $H_{\bar{x}_2}^2$ at $(\bar{x}, f(\bar{x}))$. Since these cannot coincide, the tangent cone is a hyperplane. Thus C_b is differentiable at all points $(x, f(x))$ for which $x \notin N_t$, and these form a set of measure 0 on C_b. This result implies, of course, the same result for complete surfaces.

The principal subject of this section is the distribution of the curvatures at b of the curves on a hypersurface passing through, or emanating from, a point b where the hypersurface is differentiable. This requires an agreement on the meaning of various familiar concepts under the general conditions, which interest us here.

Let $A : p(t)$, $0 \leqq t \leqq \alpha$, $\alpha > 0$, be a Jordan arc in E^n, i.e., the point $p(t)$ depends continuously on t and $p(t_1) \neq p(t_2)$ for $t_1 \neq t_2$ and put $p(0) = p$. We assume that A has at p the *tangent* T.

$$T = \lim_{t \to 0} L(p, p(t)).$$

If the (two-dimensional) plane V_t through T and $p(t)$ can be chosen (there is a choice only when $p(t) \in T$) such that it converges for $t \to 0$ to a plane V, then V is an *osculating plane* of A *at* p. If the subarc $0 \leqq t \leqq \alpha'$, $\alpha' > 0$, of A is the segment $E(p, p(\alpha'))$ then, according to this definition, any plane containing $E(p, p(\alpha'))$ is an osculating plane of A at p. We denote by $\varrho(t)$ the radius of the circle in V_t which is tangent to T at p and passes through $p(t)$. We see from (2.3) for $m = 0$ that

$$\varrho_t = | p - p(t)|^2/2 \,| p(t) - q_t |,$$

where q_t is the projection of $p(t)$ on T. Since

$$| p - p(t)|^2 = | p - q_t |^2 + | q_t - p(t)|^2$$

and

$$| q_t - p(t)| \cdot |p - q_t |^{-1} \to 0$$

we find

$$\begin{aligned} \varrho_s &= \limsup_{t \to 0} \varrho(t) = \limsup_{t \to 0} |p - q_t|^2/2 \,|p(t) - q_t| \\ \varrho_i &= \liminf_{t \to 0} \varrho(t) = \liminf_{t \to 0} |p - q_t|^2/2 \,|p(t) - q_t|. \end{aligned} \tag{3.2}$$

These numbers (∞ admitted) are *the upper and lower radii of curvature, and when equal the radius of curvature, of* A *at* p, and ϱ_s^{-1}, ϱ_i^{-1} ($= \infty$ for $\varrho_i = 0$) are the lower and upper curvatures.

If $p(t)$, $| t | \leqq \alpha$ is a Jordan arc B in E^n with $p(0) = p$ then the concepts just introduced applied to the subarc $0 \leqq t \leqq \alpha$ yield the *right tangent*, etc., of B at p. In defining tangents we must be careful to avoid cusps: T is the *tangent of* B *at* p if $T = \lim_{t \to 0} L(p, p(t))$ and $\lim_{t \to 0+} L^+(p, p(t)) = \lim_{t \to 0-} L^+(p(t), p)$. No such difficulties occur in the definition of osculating planes. Finally, B has at p a curvature if the right, left, upper, and lower curvatures coincide.

We now come to Meusnier's theorem which reduces the curvatures of curves on a hypersurface to those of normal plane sections. There are several forms of this theorem which are equivalent for smooth surfaces, but not for general surfaces. One of these is found in Bouligand [1]. For convex surfaces there is just one natural form (because they satisfy (1.9)) which was given in B. F. [1], but goes essentially back to Hjelmslev [1] who assumed a continuous tangent plane.

Consider a, not necessarily convex, hypersurface C which possesses at the point b a tangent hyperplane P in the strong sense that it contains the limit of any converging sequence of lines $L(a_\nu^1, a_\nu^2)$ through two distinct points a_ν^1, a_ν^2 on C tending to b. If N is the normal to P at b, then we know from Section 1, that with rectangular coordinates $x_1, \ldots, x_{n-1}, z$ with P as $z = 0$ and N as z-axis, a closed neighborhood $\bar{C}_b$ of b on C is representable in the form

$$z = f(x_1, \ldots, x_{n-1}), \ |x| \leqq \delta,$$

and that $|f(x^\nu) - f(y^\nu)| \cdot |x^\nu - y^\nu|^{-1} \to 0$ for $x^\nu \neq y^\nu$ and $x^\nu \to 0$, $y^\nu \to 0$. Under this hypothesis we prove:

(3.3) MEUSNIER'S THEOREM. *Let the Jordan arc $A : p(t), 0 \leqq t \leqq \alpha$, $p(0) = b$, lie on C, and have at b a tangent T and an osculating plane V which does not lie in the tangent hyperplane P of C at b. If the half plane bounded by the normal N to C at b and containing $R = \lim_{t \to 0} R(b, p(t))$ intersects $\bar{C}_b$ in the arc B then the upper and lower radii of curvature ϱ_s, ϱ_i* of A *and ϱ_s^n, ϱ_i^n of B at b satisfy*

$$\varrho_s = \varrho_s^n \cos\beta, \ \varrho_i = \varrho_i^n \cos\beta,$$

where β is the acute angle between V and N.

For a proof we let R be the positive x_1-axis, denote by p_α and p_α^n the intersections of A and B with $x_1 = \alpha$, $0 < \alpha < \delta$.[3] Then $q_\alpha = (\alpha, 0, \ldots, 0)$ is the projection of p_α and of p_α^n on T, hence by (3.2)

[3] The point p_α may not be unique, but we use only that $p_\alpha \to b$ for $\alpha \to 0+$.

$$\varrho_s = \lim\sup \alpha^2/2 \,|\, p_\alpha - q_\alpha \,|,\ \varrho_s^n = \lim\sup \alpha^2/2 \,|\, p_\alpha^n - q_\alpha \,|$$

and similarly for ϱ_i and ϱ_i^n. According to our definition of osculating plane there is a plane V_α containing T and p_α which tends for $\alpha \to 0+$ to V. If β_α is the acute angle between V_α and N, then $\beta_\alpha \to \beta \neq \frac{\pi}{2}$. If z_α, z_α^n are the z-coordinates of p_α and p_α^n then

$$|z_\alpha| = |p_\alpha - q_\alpha| \cos\beta_\alpha,\ |z_\alpha^n| = |p_\alpha^n - q_\alpha|,$$

so that the assertion will follow from $z_\alpha^n/z_\alpha \to 1$. Here we disregard the possibility that $z_{\alpha_\nu} = 0$ for a sequence $\alpha_\nu \to 0+$ which is easily taken care of, because $\beta \neq \frac{\pi}{2}$ implies that $p_{\alpha_\nu} \in T$, hence $p_{\alpha_\nu} = p_{\alpha_\nu}^n$ for large ν.

Now, if r_α is the projection of p_α on $z = 0$ then

$$|r_\alpha - q_\alpha| = \pm z_\alpha \tan\beta_\alpha,$$

and our principal hypothesis yields (because q_α is the projection of p_α^n) that

$$0 = \lim_{\alpha\to 0,\, r_\alpha \neq q_\alpha} (z_\alpha - z_\alpha^n)/\,|\, r_\alpha - q_\alpha \,| = \lim\, (z_\alpha - z_\alpha^n)/z_\alpha \tan\beta_\alpha$$

Since $\beta \neq \frac{\pi}{2}$ and $z_\alpha = z_\alpha^n$ if $r_\alpha = q_\alpha$ this implies $z_\alpha^n/z_\alpha \to 1$.

We notice the important corollary:

(3.4) *If some arc A on C emanating from, or passing through, b with tangent T at b possesses at b an osculating plane not in P and a (one- or two-sided) curvature, then the curvature exists at b for any arc on C from or through b, which has at b the tangent T and an osculating plane not in P.*

Keeping the same notations, we take the $(n-2)$-sphere $S : P \cap S(b, 1)$ as an auxiliary parameter surface whose variable point we call u. For a given ray $R(b, u)$ we define $\varrho_s(u)$ and $\varrho_i(u)$ as the upper and lower radii of curvature of the intersection of C with the half plane bounded by N and containing $R(b, u)$, and denote by $D_s(u)$, $D_i(u)$ the sets of those points x for which

$$D_s(u) : |b - x| \leqq \sqrt{\varrho_s(u)}, \quad x \in R(b, u).$$
$$D_i(u) : |b - x| \leqq \sqrt{\varrho_i(u)}, \quad x \in R(b, u).$$

If $D_s(u)$ or $D_i(u)$ is not the whole ray $R(b, u)$ we denote its end point by $r_s(u)$ or $r_i(u)$. Guided by the Dupin indicatrix of ordinary differential geometry, we say that the set of points x in P for which

$$(3.5) \quad \begin{aligned} &J_s : x \in D_s(u), \ u \in S \ \textit{is the upper indicatrix of } C \textit{ at } b. \\ &J_i : x \in D_i(u), \ u \in S \ \textit{is the lower indicatrix of } C \textit{ at } b. \end{aligned}$$

We use these sets rather than their boundaries because this makes it easier to handle the u for which $\varrho_s(u) = 0$ or $\varrho_i(u) = \infty$. Clearly $J_s \supset J_i$; if the two sets coincide we speak of the *indicatrix* $J = J_s = J_i$.

The indicatrices do not have many interesting properties for general surfaces, but they do for convex surfaces. Therefore we assume now that C is convex, and is locally represented as $z = f(x)$ with $f(x) \geqq 0$ and $z = 0$ as tangent plane at b.

For small $\epsilon > 0$ denote by J'_ϵ the intersection of the convex set $|x| \leqq \delta$, $z \geqq f(x)$ with the hyperplane $z = \epsilon/2$, and by J_ϵ the set obtained from J'_ϵ by projecting J'_ϵ on the plane $z = 0$ and then dilating it in the ratio $1 : \sqrt{\epsilon}$. The points x in J_ϵ are given by

$$J_\epsilon \colon f(\sqrt{\epsilon}\, x) \leqq \epsilon/2, \ |x| \leqq |\delta| \epsilon^{-1/2}, \ x \in P.$$

We claim that

$$(3.6) \qquad J_i = \bigcup_n \bigcap_{\epsilon < n^{-1}} J_\epsilon, \ J_s = \bigcap_n \bigcup_{\epsilon < n^{-1}} J_\epsilon$$

For a proof we put

$$M_n = \bigcap_{\epsilon < n^{-1}} J_\epsilon, \ M^n = \bigcup_{\epsilon < n^{-1}} J_\epsilon, \text{ then } M_n \subset M_{n+1}, \ M^n \supset M^{n+1}$$

$x \in M^n$ $(x \neq b)$ means that for a suitable $\epsilon_n < n^{-1}$

$$f(\sqrt{\epsilon_n}\, x) \leqq \epsilon_n/2 \ \text{ or } \ |x|^2 \leqq \epsilon_n |x|^2/2f(\sqrt{\epsilon_n}\, x)$$

It follows that for $x = \alpha u \in \bigcap M_n$, $\alpha > 0$,

$$\varrho_s(u) \geqq \limsup_{n \to \infty} \epsilon_n |x|^2/2 f(\sqrt{\epsilon_n}\, x) \geqq |x|^2$$

or $x \in J_s$ and $J_s \subset \bigcap M^n$.

Conversely, if $x \in J_s$, $x = \alpha u$, $\alpha > 0$, then $x \in \cap M^n$ is trivial for $\varrho_s(u) = \infty$. If $\varrho_s(u)$ is finite then $\epsilon_n < n^{-1}$ with

$$\epsilon_n \mid x \mid^2/2 \, f(\sqrt{\epsilon_n}\, x) < \varrho_s(u) + 1/n$$

exists, so that M^n lies in the set obtained from J_s by prolonging each finite $D_s(u)$ by a segment of length $1/n$; therefore $\cap M^n \subset J_s$.

Similarly, if $x \in M_n$ then for all $\epsilon < n^{-1}$

$$\epsilon \mid x \mid^2/2 \, f(\sqrt{\epsilon}\, x) \geqq \mid x \mid^2$$

so that with $x = \alpha u$, $\alpha > 0$,

$$\varrho_i(u) = \liminf_{\epsilon \to 0+} \epsilon \mid x \mid^2/2 \; f(\sqrt{\epsilon}\, x) \geqq \mid x \mid^2$$

or $M_n \subset J_i$ and $\cup M_n \subset J_i$. If $x \in J_i$ then

$$\epsilon \mid x \mid^2/2 \; f(\sqrt{\epsilon}\, x) < \varrho_i(u) + \eta_n(u) \text{ for } \epsilon < n^{-1}$$

with $\eta_n(u) \to 0$, or M_n lies in the set obtained from J_i by prolonging each finite $D_i(u)$ by $\eta_n(u)$. Since $M_n \subset M_{n+1}$ the relation $\cup M_n \subset J_i$ follows.

Since J_ϵ is convex and $M_n \subset M_{n+1}$, the lower indicatrix J_i is a convex set. Also, if $x \in M^n$, then $x \in J_\epsilon$ with a suitable $\epsilon < n^{-1}$ and $J_\epsilon \supset M_n$, hence M^n contains all segments $E(x, y)$ with $y \in M_n$. For $x \in J_s$ we find that J_s contains all segments $E(x, y)$ with $y \in \cup M_n = J_i$. This is the smallest convex set containing $x \cup J_i$ or, according to standard terminology, the convex closure of x and J_i. We have found (compare B. F. [1]):

(3.7) THEOREM. *The lower indicatrix J_i of a convex hypersurface at a point b (where it is differentiable) is convex. The upper indicatrix J_s contains with any point x the convex closure of $x \cup J_i$.*

If b is an interior point of J_i then both $\varrho_i(u)$ and $\varrho_s(u)$ are continuous functions of u.

The continuity of $\sqrt{\varrho_i(u)}$ was seen in the first section. The fact that J_s contains with the boundary point $u\sqrt{\varrho_i(u)}$ the convex closure of this point and a ball $U(p, \beta)$ with $\beta > 0$ shows readily that $\sqrt{\varrho_s(u)}$ is a continuous function of u.

It is easy to construct a hypersurface which has a given convex closed set J in P, containing b in the interior, as indicatrix: If with a sort of cylindrical coordinates ϱ, u, z the boundary of J has the equation $\varrho = \lambda(u)$, where $\lambda(u) = \infty$ if $R(p, u) \subset J$, then

$$z = \varrho^2/2\lambda^2(u)$$

is such a hypersurface.

We are principally interested in the case where $J_i = J_s$ contains b in the interior and therefore merely report without proof that, for $n = 3$, according to B. F. [2, II], the conditions of (3.7) are,

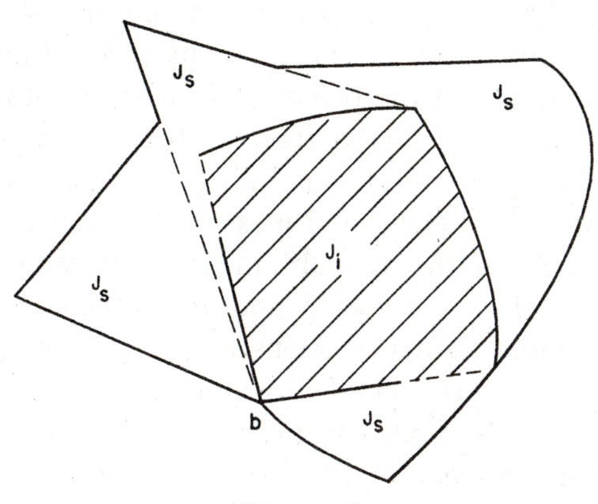

Figure 2

unless $J_i = b$, not only necessary, but also *sufficient* for J_s and J_i to be the upper and lower indicatrices at a point of a suitable convex surface. Some surprising possibilities occur; for example, if an angle φ replaces u, then we may have $\varrho_i(\varphi) = 1$ for $0 < \varphi < \beta \leqq \pi$, $\varrho(\varphi) = 0$ for $\beta < \varphi < 2\pi$ and arbitrary numbers in $[0,1]$ as $\varrho_i(0)$ and $\varrho_i(\beta)$. For $J_i = b$ the condition on $\varrho_s(u)$ in (3.7) is empty. In that case it is necessary and sufficient for $\varrho_s(\varphi)$ to represent the upper radius of curvature of a suitable convex surface that $\varrho_s(\varphi)$ be the limit of a decreasing sequence of lower semicontinuous functions. The case where $J_i = J_s = J$ contains b in the interior is the most interesting because of:

(3.8) *At almost all points b of a convex hypersurface the indicatrix exists and has b as interior point and center.*

The theorem holds for $n = 2$, because the points of the convex curve $y = f(x)$ with $|f'_m(x)| < \infty$ where a finite two-sided curvature or a positive radius of curvature exists are, according to Section 2, exactly those where $f''(x)$ exists and is finite; and this is the case for almost all x.

The case $n = 2$ yields the theorem for $n = 3$: If, with the previous notations, C_b is given by $z = f(x_1, x_2)$, then the plane $x_2 = mx_1 + \beta$ will intersect C_b in a curve B_β which has almost everywhere a curvature. At a point q of B_β where C_b has a tangent plane and B_β has a finite curvature, the normal plane section of C_b through q and the tangent of B_β has by (3.4) a finite curvature. Varying β we find that the points x in the z-plane for which C_b is not differentiable at $q_x = (x, f(x))$ or the normal section of C_b through the tangent of B_β through q_x does not have a finite curvature, has two-dimensional measure 0, hence the same holds for the union N_j of the sets obtained in the same way for the different rational m. If $x \notin N_j$, and P_x is the tangent plane of C_b at q_x, then the normal sections of C_b through q_x have finite curvatures for a dense set of directions in P_x. The continuity properties of $\varrho_i(u)$ and $\varrho_s(u)$ in (3.7) imply that $J_i = J_s$, and the finiteness and existence of the two-sided curvatures mean that J_i has q_x as interior point and center.

Clearly, the same type of argument yields an inductive proof for general n.

Surprisingly enough, except for a new set of measure 0, Euler's theorem holds in the classical sense, i.e., the indicatrix J is bounded by a quadric, which may, of course, degenerate in various ways; for example, if $n = 4$, and the indicatrix J at b lies in an E^3, then its boundary may be: an ellipsoid with center b, an elliptic cylinder with center b, a pair of parallel planes with equal distance from b, and finally J may be the whole E^3 when all normal curvatures vanish.

The decisive step in the proof of Euler's theorem is evidently the case $n = 3$ treated in B. F. [1]. For then the same argument as above shows that for $n = 4$ at almost all points q_x whose projections do not lie in N_j the boundary of the indicatrix J is intersected in conics by a set of planes through b which is dense among all planes through b; and hence must be a quadric.

Therefore, we consider a convex surface C_b represented in the form

$$z = f(x_1, x_2), \ |x| < \delta,$$

with bounded $(f(x) - f(y)) \, |x - y|^{-1}$.

For each interval

$$Q : x_i - h_i \leqq x \leqq x_i + h_i, \ h_i > 0 \text{ in } |x| < \delta$$

we put

$$F(Q) = \Delta(x, h) = f(x_1 + h_1, x_2 + h_2) + f(x_1 - h_1, x_2 - h_2) \\ - f(x_1 - h_1, x_2 + h_2) - f(x_1 + h_1, x_2 - h_2).$$

The convexity of $f(x)$ entails the existence of a constant M such that for any finite set $Q_1, \ldots, Q_n$ of non-overlapping intervals in $|x| < \delta$

$$\sum_{\nu=1}^{n} |F(Q_\nu)| < M,$$

i.e., $f(x)$ has bounded variation in the sense of Caratheodory. A proof for a much more general class of surfaces is found in B. F. [3], pp. 592, 593. According to well-known results in real variable theory the limit

$$\text{(3.9)} \qquad \lim_{\nu \to \infty} \frac{\Delta(x, h^\nu)}{4 \, h_1^\nu, h_2^\nu} = D(x),$$

when $h_i^\nu \to 0+$ and $0 < \lim\inf h_1^\nu / h_2^\nu < \lim\sup h_1^\nu / h_2^\nu < \infty$ exists on a set N' of measure 0 and is independent of the choice of the sequence h^ν (see for example Hahn and Rosenthal [1], Section 19).

Consider an (x_1, x_2) which does not lie in $N_j \cup N'$. Then for $h_\alpha =$ $(h \cos\alpha, h \sin\alpha)$, $0 < \alpha < \pi/2$,

$$D(x) = \lim_{h \to 0} \Delta(x, h_\alpha)/4h^2 \cos\alpha \sin\alpha.$$

Since $x \notin N_j$, and because of (2.2), the directional derivative

$$f_{\alpha\alpha}(x) = \lim h^{-2} \, f(x_1 + h\cos\alpha, x_2 + h\sin\alpha) \\ + f(x_1 - h\cos\alpha, x_2 - h\sin\alpha) - 2f(x_1, x_2)$$

exists for each α. But

$$\Delta(x, h_\alpha) = f(x_1 + h \cos \alpha, x_2 + h \sin \alpha) \\ + f(x_1 - h \cos \alpha, x_2 - h \sin \alpha) - 2f(x_1, x_2) \\ -[f(x_1 + h \cos(-\alpha), x_2 + h \sin(-\alpha) + \\ f(x_1 - h \cos(-\alpha), x_2 - h \sin(-\alpha)) - 2f(x_1, x_2)]$$

so that

$$f_{\alpha\alpha}(x) - f_{-\alpha,-\alpha}(x) = 4 \sin \alpha \cos \alpha D(x).$$

We now rotate the coordinate system about the z-axis through an angle ω with irrational ω/π. If (v_1, v_2) are the new coordinates and $f(x)$ becomes $g(v_1, v_2)$, then the set N_j is the same for the new system, but the set where the derivative $D(x)$ analogous to (3.9) does not exist will, in general, be a set $N'' \neq N'$. As before we obtain for $v \notin N_j \cup N''$

$$g_{\beta\beta}(v) - g_{-\beta,-\beta}(v) = 4 \sin \beta \cos \beta D^*(v).$$

If we transform this relation back into the old coordinates we obtain

$$f_{\alpha\alpha}(x) - f_{2\omega-\alpha,2\omega-\alpha}(x) = 4 \sin(\alpha - \omega) \cos(\alpha - \omega) D^*(v).$$

The function $F(\alpha) = f_{\alpha\alpha}(x)$ is by (3.7) a continuous function of α and satisfies for $x \notin N_j \cup N' \cup N''$ two functional equations of the form

$$F(\alpha) - F(-\alpha) = 2a \sin 2\alpha$$
$$F(\alpha) - F(2\omega - \alpha) = 2b \sin 2(\alpha - \omega),$$

which have the solution

$$F(\alpha) = k \sin 2\alpha + k' \cos 2\alpha + k''$$

with $k = a$, $k' = (a \cos 2\omega - b)/\sin 2\omega$, $k'' = F(0) - k'$. The functional equations determine $F(2n\omega)$ successively by

$$F(2\omega) = F(0) + 2b \sin 2\omega,$$
$$F(2(n+1)\omega) = F(-2n\omega) + 2b \sin 2(2n+1)\omega.$$

Because ω/π is irrational the angles $2n\omega$ are dense, therefore the functional equations have one continuous solution for given $F(0)$. Thus

$$f_{\alpha\alpha}(x) = k_1 \sin^2 \alpha + k_2 \cos^2 \alpha + k_3 \sin \alpha \cos \alpha$$

with coefficients which depend only on x. If the tangent plane at x_0 is parallel to $z = 0$, then this is Euler's theorem, because $f_{\alpha\alpha}(x_0)$ is the curvature of the normal section $(x - x_0) \cos \alpha = (y - y_0) \sin \alpha$. In the general case, Meusnier's theorem gives the relation between $f_{\alpha\alpha}(x_0)$ and the curvature of the normal section through the tangent at x_0 of the section $(x-x_0) \cos \alpha = (y-y_0) \sin \alpha$ and this yields Euler's theorem. The simple calculation is omitted here, but may be found in B. F. [1], p 29. Thus we have proved:

(3.10) THEOREM. *An arbitrary convex hypersurface satisfies almost everywhere Euler's theorem, i.e., the following statements hold simultaneously almost everywhere on the hypersurface: the tangent plane exists, all normal plane sections have two-sided finite curvatures, the boundary of the indicatrix is a quadric.*

As for curves the question arises whether this theorem holds for a wider class of surfaces. The following is proved in B.F. [3]:

Let the surface $x(u) = (x_1(u_1, u_2), x_2(u_1, u_2), x_3(u_1, u_2))$ in E^3 have the following properties: For a given u^0 there is 1) a $\delta(u^0) > 0$ such that the mapping $x(u) \to u$ is topological for the subsurface S_0: $x(u_1, u_2)$, $|u_i - u_i^0| \leqq \delta(u^0)$. 2) A line L through $x(u^0)$ exists which is not the limit of lines $L(x(u^\nu), x(v^\nu))$ with $u^\nu \neq v^\nu$ and $u^\nu \to u^0$ $v^\nu \to u^0$. 3) The planes parallel to L intersect S_0 in curves with uniformly bounded total curvatures (see end of Section 2). Then Euler's theorem holds almost everywhere on the surface.

This class of surfaces is closely related to the surfaces of bounded extrinsic curvature which are investigated in Pogorelov [9].

Consider a point b of a convex hypersurface where the surface is differentiable and has $z = 0$ as tangent plane and the indicatrix J exists and contains b in the interior; with the earlier notations put

$$g(u) = \varrho_i^{-1}(u) = \varrho_s^{-1}(u) > 0.$$

Then $J_\epsilon \to J$ shows that for $x_n = t_n u_n \to x = tu$

$$\lim \frac{2f(x_n)}{|x_n|^2} = g(u)$$

hence

$$f(x) = \frac{|x|^2}{2} g(u) + o(|x|^2)$$

If Euler's theorem holds at b this may be written in the form

$$f(x) = \tfrac{1}{2} \sum x_i x_j a_{ij} + o(|x|^2), \; a_{ij} = a_{ji}.$$

If $(q, f(q))$ is any point of the hypersurface where Euler's theorem holds (and the tangent plane is not parallel to the z-axis)this relation takes the form

$$f(x) = f(q) + \sum_i (x_i - q_i) \frac{\partial f(q)}{\partial x_i}$$
$$+ \tfrac{1}{2} \sum_{i,j} (x_i - q_i)(x_j - q_j) b_{ij} + o(|x|^2), \; b_{ij} = b_{ji},$$

so that $f(x)$ has a sort of generalized second differential at q. Alexandrov showed in [2] that this is actually an *ordinary second differential* in the following sense:

Similarly as in Section 2, we denote by $\partial_m f(x)/\partial x_i$ the slope of any supporting line of the curve $x_\nu = q_\nu$, $\nu \neq i$ at $(x, f(x))$, then b_{ij} is the second partial derivative

$$b_{ij} = \lim_{h \to 0} h^{-1} \left(\frac{\partial_m f(q_1, \ldots, q_{j-1}, q_j + h, q_{j+1}, \ldots, q_{n-1})}{\partial x_i} - \frac{\partial f(q)}{\partial x_i} \right),$$

and is, because $b_{ij} = b_{ji}$, independent of the order of the differentiations. He also proved the deeper fact that at a point where Euler's theorem holds the *directional derivatives* of $\partial f/\partial x_i$ *are uniformly approximated by the corresponding difference quotients*:

$$\left| \frac{1}{|x|} \left(\frac{\partial_m f(q+x)}{\partial x_i} - \frac{\partial f(q)}{\partial x_i} \right) - \lim_{h \to 0+} \frac{1}{h|x|} \left(\frac{\partial_m f(q+hx)}{\partial x_i} - \frac{\partial f(q)}{\partial x_i} \right) \right| = o\,(|x|)$$

4. Extrinsic Gauss curvature

Henceforth we denote the origin by z and the unit sphere $S(z, 1)$ by Z. A set M on Z is convex if it contains with any two points u, v at least one shortest connection of u and v, i.e., either a semi great circle from u to v if u and v are antipodes, or, if they are not, the shorter arc of the great circle through u and v.

(4.1) *A convex set M on Z either is contained in a closed hemisphere of Z or $M = Z$.*

For, owing to the convexity of M, the union of all rays $R(z, u)$, $u \in M$, is a convex set M' in E^n. If z is an interior point of M', then $M' = E^n$ and $M = Z$. If z is a boundary point of M then M' possesses at z a supporting plane whose intersection with Z bounds a closed hemisphere containing M.

Consider a complete convex hypersurface C which is not a hyperplane and hence bounds exactly one convex set K. We orient the unit normal u of a given supporting plane P of C such that it becomes the exterior normal, i.e., the end point of a vector parallel to u beginning at a point of P lies on the other side of P from K. For a given u there is at most one supporting plane P_u of C with normal u.

The *spherical image* $\nu'(p)$ of a point $p \in C$ consists of the end points of all unit vectors with origin z which are normal to supporting planes of C at p. The spherical image $\nu'(M)$ of a set $M C C$ is

$$\nu'(M) = \bigcup_{p \in M} \nu'(p)$$

(4.2) *$\nu'(M)$ is closed if M is closed.*

(4.3) *The image $\nu'(p)$ of a point p is convex.*

This is equivalent to the convexity of the polar cone T' to the tangent cone T of C at p which we already used in the discussion preceding (1.12).

(4.4) *$\nu'(C)$ is convex, and hence coincides with Z or lies on a closed hemisphere of Z.*

If $n = 2$, then $\nu'(C)$ is a connected subset of the unit circle. If $\nu'(C)$ properly contains a closed semicircle then C has two distinct pairs of parallel supporting lines, hence lies in a parallelogram and is closed so that $\nu'(C) = Z$.

If $n > 2$ and u, v are two non-antipodal points of $\nu'(C)$ then P_u and P_v intersect in an $(n-2)$-flat V. The intersection of all supporting half spaces of K bounded by supporting planes of C parallel to V is a convex set. It intersects the two-flat V' through z, u, v in a convex set. The boundary of this set in V' is a convex curve C' whose spherical image on $V' \cap Z$ contains the shorter arc A of the circle of $V' \cap Z$ from u to v. Clearly $A \subset \nu'(C)$. When u and w are antipodal points in $\nu'(C)$ we take any third point $v \in \nu'(C)$ (which exists because C is not a pair of parallel hyperplanes) and proceed with u, v and with v, w as before. The spherical image of C contains then a semi great circle from u to w.

A complete open convex hypersurface C is representable in the form $x_n = f(x_1, \ldots, x_{n-1}) \geqq 0$ if and only if $\nu'(C)$ lies in the open hemisphere with center $(0, \ldots, 0, -1)$.
For such a representation exists if and only if C has no supporting line parallel to the x_n-axis.

We put for any set M of C for which $\nu'(M)$ is measurable,

$$\nu(M) = \textit{measure of } \nu'(M) = \textit{integral curvature of } M.$$

Because of the convexity of $\nu'(C)$ this set lies in a hyperplane if and only if $\nu(C) = 0$.

The problems which will interest us in the remainder of this section become either trivial or uninteresting if $\nu(C) = 0$. We therefore assume $\nu(C) > 0$. Then by (4.4) with

(4.5) $\pi_n = \pi^{n/2}/\Gamma(n/2+1)$ *hence volume* $\overline{U}(z, 1) = \pi_n$, *area* $Z = n\pi_n$,

C is closed and $\nu(C) = n\pi_n$ or C is open and $0 < \nu(C) \leqq n\pi_n/2$. Starting from (4.2) one proves

(4.6) THEOREM. *$\nu'(M)$ is measurable for any Borel set M on C and the integral curvature $\nu(M)$ is completely additive on the Borel sets of C.*

This result is due to Alexandrov. A detailed proof for $n = 3$ (which extends to arbitrary n, see A. [2]) is found in A., pp. 186–195. In the main it uses well-known arguments from real variable theory, but there is one difficulty peculiar to the present case: $M_1 \cap M_2 = 0$ does not imply $\nu'(M_1) \cap \nu'(M_2) = 0$; it implies however that the latter set has measure 0, because the set of those $u \in Z$, for which P_u contains more than one point of C, has measure 0.

Many interesting questions now arise. It will be seen later, in Section 14, that $\nu(M)$ is for $n = 3$ an intrinsic quantity. If Euler's theorem holds at b, we call b briefly an *Euler point* and denote by $K_E(b)$ the usual expression for the curvature, namely, if the indicatrix J is an ellipsoid the product of the squares of the reciprocals of the semiaxes and 0 if J is unbounded.

The principal curvatures can under very general conditions be given the familiar interpretation known as the theorem of Olindes-Rodrigues: Assume that C is differentiable and strictly convex in a neighborhood of b (so that the spherical mapping is topological) and that the indicatrix J exists, is differentiable and contains b in its interior. If for $q \in J$ the ray $R(b, q)$ is normal to the tangent plane of J at q, consider a rectifiable curve $x(s)$, $0 \leqq s \leqq s_0$, $x(0) = b$, on C which has $R(b, q)$ as right tangent at b and has a finite upper curvature at b. Then the spherical image $z(s)$ of $x(s)$ satisfies

$$\frac{dz(0)}{ds} = \frac{1}{\varrho} \frac{dx(0)}{ds}$$

where $\varrho = | b - q |^2$ is the normal curvature of C in the direction of $R(b, q)$. This result remains (in a trivial way) true for directions where $\varrho = \infty$. A proof for $n = 3$ is found in B. F. [1]; it can be extended to $n > 3$.

Denote by $A(M)$ the measure (or area) of the Borel set M on C. In analogy to specific weight the ratio $\nu(M)/A(M)$ is called the *specific curvature* of M. The next question concerns the limit of this specific curvature if M shrinks to a point b.

Even with a reasonable method of shrinking M to b the specific curvature of M may not converge at Euler points and may exist at points which are not. This is confirmed by the following results found in B. F. [2, III]: For $n = 3$ consider a point b of a convex surface C where C has an indicatrix J. With the notations of the previous section let (with φ for u) $\varrho(\varphi) = \varrho_i(\varphi) = \varrho_s(\varphi)$ be the radius of curvature in the direction φ at b and denote by C_h the cap cut off from C by a plane parallel to, and at distance $h > 0$ from, the tangent plane of C at b. If $\varrho(\varphi)$ is bounded and takes the value 0, (i.e., b lies on the boundary of J) then

$$\lim_{h \to 0+} \nu(C_h)/A(C_h) = \infty.$$

If $0 < \varrho(\varphi) < \infty$ then

$$\lim_{h \to 0+} \nu(C_h)/A(C_h) = \int_0^{2\pi} [\varrho^2(\varphi) - \varrho'^2(\varphi)/4]\varrho^{-3}(\varphi)d\varphi \Big/ \int_0^{2\pi} \varrho(\varphi)d\varphi.$$

This becomes $K_E(b)$ if b is an Euler point. If $\varrho(\varphi)$ is not bounded then $\lim \nu(C_h)/A(C_h)$ need not exist even if b is an Euler point.

From the intrinsic point of view, geodesic circular disks are preferable to caps: the intrinsic distance pq of two points on a convex hypersurface C is the length of a shortest connection of p and q on C (for greater details see Section 11). The disk or, in arbitrary dimensions, the ball D_h about b with radius h consists of the points on C with $bp \leqq h$. Since Vitali's covering theorem holds for balls, real variable theory (see Hahn-Rosenthal [1], Section 19) suggests considering Borel sets M_i on C which tend to b in such a way that for the smallest ball D_{h_i} which contains M_i the ratio

(4.7) $$A(D_{h_i})/A(M_i) \text{ is bounded.}$$

It is shown in B. F. [2, III] that *for almost all Euler points on a convex surface*

(4.8) $$K_E(b) = \lim_i \nu(M_i)/A(M_i).$$

Under a condition which is slightly more stringent than (4.7), Alexandrov proves in [2] that (4.8) holds at *every* Euler point of a convex hypersurface.

The set function $\nu(M)$ is, by definition, absolutely continuous if $A(M) = 0$ implies $\nu(M) = 0$. Standard real variable theory, (3.10) and (4.8) yield:

(4.9) THEOREM. *If $\nu(M)$ is absolutely continuous, then for any Borel set M on a convex hypersurface*

$$\nu(M) = \int_{M_E} K_E(b)dA,$$

where M_E consists of the Euler points of M.

If $\lim_{h \to 0} \nu(D_h)/A(D_h)$ exists for every $b \in C$ and is a continuous function of b, then the set D_h may be replaced by arbitrary Borel sets M_i with $A(M_i) > 0$ tending to b without changing the limit. Even then it does not follow that all points are Euler points.

$$2x_3 = x_1^2 \log x_3 - x_2^2/\log x_3, \quad x_3^2 = |x_1|^7 + |x_2|^3$$

are examples of convex surfaces in E^3 where $\lim \nu(D_h)/A(D_h)$ exists everywhere and is continuous. But the origin z is no Euler point. In the first example there is not even an indicatrix at z and the limit has value 1; in the second, the indicatrix is the x_1-axis and the limit has value 0.

Further problems regarding $\nu(M)$ concern the existence and uniqueness of a convex surface with a given integral curvature $\nu(M)$. The most natural formulation is Minkowski's problem which gives the measure $A(\nu')$ of the inverse image M on the surface of the set ν on Z under the spherical mapping. It is the subject of Section 8.

For the corresponding problem regarding $\nu(M)$ place the origin z inside the closed convex surface C. Denote by M' the projection from z on Z of the set $M \subset C$ and put $\alpha(M') = \nu(M)$. Alexandrov proved in [3], see also A. [8], Chapter IX:

(4.10) THEOREM. *For a given completely additive non-negative set function $\alpha(M')$ defined on all Borel sets $M' \subset Z$ there exists a closed convex hypersurface C containing z in its interior such that $\alpha(M') = \nu(M)$ for the projection M of M' from z on C, if and only if 1) $\alpha(Z) = n\pi_n$ and 2) $\alpha(K) \leq n\pi_n - \beta$ for every convex set K on Z, where β is the measure of the spherical image of the cone projecting K from z.*

The necessity of the conditions is obvious. We omit the sufficiency proof because it is very similar to (in fact, simpler than) the existence proof in Minkowski's problem. It also proceeds by approximation with polyhedra. The corresponding uniqueness theorem may be formulated as follows (see A. [4]).

(4.11) THEOREM. *Let C_1 and C_2 be two closed convex hypersurfaces containing z in the interior. If $\nu(M_1) = \nu(M_2)$ for any two Borel sets $M_j \subset C_j$ which are projections of each other from z, then C_2 is obtained from C_1 by a dilation with center z.*

We give the proof because it is essentially different from the uniqueness proof in Minkowski's problem. We call two points and sets on C_1 and C_2 corresponding if one is obtained from the other by projection from z. The differentiability properties of convex hypersurfaces yield readily that C_2 can be obtained from C_1 by a dilation with center z if C_1 and C_2 possess parallel tangent planes at all pairs of corresponding points where both C_1 and C_2 are differentiable. Assume for an indirect proof that a pair of corresponding points $p_j \in C_j$ exist where C_1 and C_2 are differentiable but the tangent planes P_i are not parallel. By a dilation of C_2 from z we obtain a surface through p_1 which we call again C_2. Then C_1 and C_2 intersect at $p_2 = p_1$.

Put $C = C_1 \cap C_2$ and denote by

C_{11} (or C_{22}) the part of C_1 (or C_2) lying outside C_2 (or C_1)

and by

C_{12} (or C_{21}) the part of C_1 (or C_2) lying inside C_2 (or C_1).

The sets C_{11} and C_{21} are projections of each other, hence by hypothesis $\nu(C_{11}) = \nu(C_{21})$. However we can show $\nu(C_{11}) > \nu(C_{21})$. In the first place $\nu'(C_{11}) \supset \nu'(C_{21})$, because a supporting plane Q of C_2 at a point of C_{21} cuts a cap off C_1 which has a supporting plane parallel to Q. The sets $C_{12} \cup C$ and $C_{21} \cup C$ are closed, hence, see (4.2), $\nu'(C_{12} \cup C)$ and $\nu'(C_{21} \cup C)$ are closed, so that $Z - \nu'(C_{12} \cup C)$ and $Z - \nu'(C_{21} \cup C)$ are open. On the other hand

$$Z - \nu'(C_{12} \cup C) \subset \nu'(C_{11}) \text{ and } [Z - \nu'(C_{21} \cup C)] \cap \nu'(C_{21}) = 0$$

Therefore

$$M = [Z - \nu'(C_{12} \cup C)] \cap [Z - \nu'(C_{21} \cup C)]$$

is open, $M \cap \nu'(C_{21}) = 0$ and $M \subset \nu'(C_{11})$. If M is not empty then $\nu(C_{11}) > \nu(C_{21})$ follows.

One, P, of the two hyperplanes through $P_1 \cap P_2$ bisecting the angle between P_1 and P_2 cuts pieces off C_{11} and C_{22}, which have supporting planes parallel to P. The spherical image of P lies neither in $\nu'(C_{12} \cup C)$ nor in $\nu'(C_{21} \cup C)$, hence lies in M.

A. [4] extends these results to open surfaces. In order to obtain formulations free from exceptions it is convenient to include — in the remainder of this section only — among the open hypersurfaces the boundary C of any unbounded convex set K in E^n which does not contain an entire line. Thus a ray and a plane convex set bounded by a curve other than a line are convex hypersurfaces. It is clear how the spherical image of a set on such a hypersurface is to be defined, and that it still lies on a closed hemisphere of Z. Hence there exists a hyperplane P such that a line perpendicular to P either does not intersect K or intersects K in a ray.

If M' is for $M \subset C$ the orthogonal projection of M on P, then $\alpha(M') = \nu(M)$ is a completely additive set function in P defined for all Borel sets in P, provided we put $\alpha(M') = 0$ for any set M' in P where the perpendicular to P at a point of M' does not intersect K. For any such function $\alpha(M')$ with $0 < \alpha(P) \leqq n\pi_n/2$ there is a convex hypersurface C with $\nu(M) = \alpha(M')$, but C will in general, not be unique. In order to obtain uniqueness as well as existence we define as *limit cone* of C a cone with the arbitrary apex a which has the same spherical image as C. It is the intersection of all half spaces whose bounding hyperplanes pass through a and are obtainable by translations from supporting half spaces of K. If $\nu(C) = n\pi_n/2$ then a limit cone is a ray. A. [4] establishes:

(4.12) THEOREM. *Let $\alpha(M')$ be a non-negative completely additive set function defined on all Borel sets of the hyperplane P such that $0 < \alpha(P) \leqq n\pi_n/2$. Let D be a convex cone with $\nu(D) = \alpha(P)$ and so situated that a line normal to P intersects the convex set bounded by D either in a ray or not at all. Then there is one, and up to translations normal to P, exactly one convex hypersurface C for which D is a limit cone and $\alpha(M') = \nu(M)$.*

The proof is analogous to the case of closed hypersurfaces. In the uniqueness proof the unboundedness of C_1 and C_2 does not prevent the use of caps because C_1 and C_2 have the same limit cone. If $\alpha(P) = n\pi_n/2$ then the limit cone with apex a is unique (namely a ray), hence D need not be mentioned in (4.11).

5. The influence of the curvature on the local shape of a surface

Our last investigations concerned the determination in the large of a hypersurface by its integral curvature. In this section we study local implications of properties of $\nu(M)$ in the case $n = 3$, a corresponding theory for $n > 3$ does not yet exist. We begin with a theorem of Alexandrov, see A. [5].

(5.1) THEOREM. *If the specific curvature of the convex surface C is bounded (on all Borel sets of positive measure on C), and C is not differentiable at $b \in C$, then b is interior point of a straight edge on C.*

A straight edge is a non-degenerate segment on C which lies in two distinct supporting planes of C.

For a proof we must exclude three eventualities:

1) b is a conical point of C, i.e., there are three independent supporting planes of C through b;

2) There are (exactly) two independent supporting planes through b, but b does not lie on a straight edge;

3) b is an end point of a straight edge, but not conical.

In case 1 we have $\nu(b) > 0$, hence $\nu(M) \geqq \nu(b)$ for any Borel set containing b, hence the specific curvature is not bounded.

In the second case the tangent cone of C at b consists of two half planes T_1, T_2 bounded by a line L through b. Consider a plane Q at distance h from, and parallel to, L which intersects T_1 and T_2 and forms equal angles with T_1 and T_2. Since b does not lie on a straight edge Q will, for small $h > 0$, cut off a cap C_h from C. Denote by P_1, P_2 the two supporting planes of C_h normal to L. If $\angle\,(T_1, T_2) = 2\alpha$, $P_i \cap L = p_i$ and $x_i = |\,b - p_i\,|$ the union D of the faces of the prism bounded by P_1, P_2, T_1, T_2 and Q which do not lie in Q, has area

$$A(D) = 2h(h \sin \alpha + x_1 + x_2)/\cos \alpha \geqq A(C_h).$$

To estimate $\nu(C_h)$ we consider the cone K obtained by connecting

b to the points of the boundary $C \cap Q$ of C_h. Then $\nu'(C_h) \supset \nu'(K)$ because for every supporting plane of K there is a parallel supporting plane of C_h (at a point not in $C \cap Q$). For the same reason $\nu'(K)$ contains the spherical image of the pyramid with apex p and

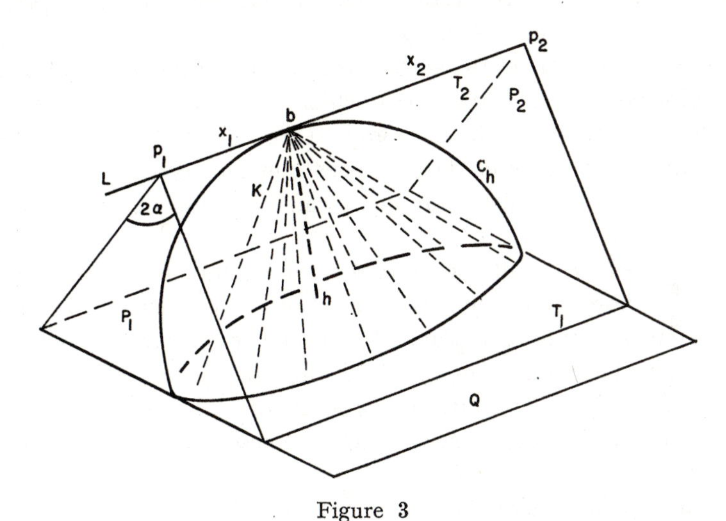

Figure 3

the face of the prism in Q as base. The spherical image of this pyramid is a spherical quadrangle whose diagonals intersect at right angles and have length $\pi - 2\alpha$ and

$$\pi - 2\beta = \text{arc tan } (h/x_1) + \text{arc tan } (h/x_2).$$

Since the spherical metric is locally euclidean, the area of the spherical quadrangle will for small h be approximately the euclidean area $(\pi - 2\alpha)(\pi - 2\beta)/2$. Hence there is a constant $\delta > 0$ such that for small h

$$\nu(C_h) \geqq \nu(K) > 4\delta(\pi - 2\alpha)\ (\pi - 2\beta) > 2\delta(\pi - 2\alpha)\ (h/x_1 + h/x_2).$$

Therefore

$$\nu(C_h)/A(C_h) > \delta \cos \alpha(\pi - 2\alpha)\ (x_1^{-1} + x_2^{-1})\ (h \sin \alpha + x_1 + x_2)^{-1}$$

which tends to ∞ for $h \to 0$ because $x_i \to 0$.

The third case is treated similarly: if b is an end point of the straight edge E we choose Q at a small distance h from b through a point of E sufficiently close to b.

This theorem has very important corollaries:

(5.2) THEOREM. *A closed convex surface with bounded specific curvature is everywhere differentiable.*

For, a point where the surface is not differentiable would be an interior point of a straight edge which, owing to the boundedness of the surface, would have to end. (We remember that differentiability implies, for convex surfaces, continuous differentiability, see (1.11)). A similar statement can be made on special non-complete surfaces.

A convex *cap* is an open subset C' of a complete convex hypersurface whose boundary B is a closed $(n-2)$-dimensional convex hypersurface in a hyperplane P and such that a normal to P at a point inside or on B intersects $\bar{C}'$ in exactly one point.

Then we have the further corollary of (5.1):

(5.3) *A two dimensional convex cap with bounded specific curvature is everywhere differentiable.*

For a point where the cap is not differentiable must lie on a straight edge with end points on the boundary of the cap. Then the cap is a plane domain.

We now prove another theorem of a similar type as (5.1), which is stated without proof in A. [5]; our proof is due to Pogorelov [3]:

(5.4) THEOREM. *If the convex surface C contains a proper segment E, then any interior point b of E lies in a Borel set M of arbitrarily small diameter whose specific curvature is arbitrarily small.*

The tangent cone of C at b consists of two half planes T_1 and T_2 bounded by the line carrying E. (Now $T_1 \cup T_2$ may be a plane). Let a_1, a_2 be points on E equidistant from b and put $\alpha = |a_1 - a_2|$.

The plane normal to E at b intersects C in a curve B. We may assume that B contains no proper segment beginning at b because in that case C contains a plane triangle and the assertion is obvious.

Take a point p with small $|p - b| = h > 0$ on the exterior normal to C at b which bisects the angle between the normals to T_1 and T_2 at b (or on the normal to $T_1 \cup T_2$ if this is a plane). Consider the polyhedral angle D_h with apex p and four faces whose edges are $R(p, a_1)$, $R(p, a_2)$ and the two rays from p on supporting lines to B. This angle or cone cuts out quadrangular regions T_h from $T_1 \cup T_2$ and C_h from C.

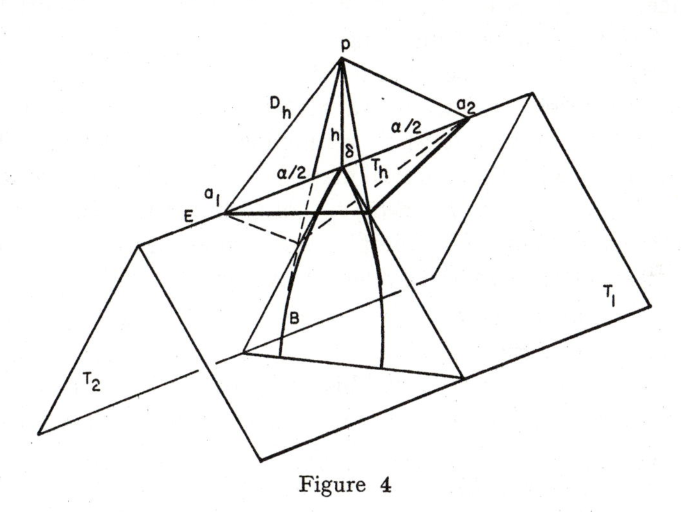

Figure 4

Clearly

$$A(C_h) \geqq A(T_h) \text{ and } \nu(D_h) \geqq \nu(C_h).$$

If h is small compared to α, then for a suitable constant $\delta > 0$

$$\nu(D_h) < \delta h/\alpha, \; A(T_h) = \alpha(\epsilon_1(h) + \epsilon_2(h))/2,$$

where $\epsilon_i(h)$ are the distances of b from the vertices of T_h different from a_1 and a_2. Hence

$$\nu(C_h)/A(C_h) \leqq \nu(D_h)/A(T_h) \leqq 2\delta h/\alpha^2(\epsilon_1(h) + \epsilon_2(h)).$$

But $h/\epsilon_i(h) \to 0$ for $h \to 0+$.

The following is an important corollary of (5.4):

(5.5) *A convex surface whose specific curvature is bounded away from 0 is strictly convex.*

Assume now that for every point x of a not necessarily complete convex surface C the limit

$$K(x) = \lim_{M \to 1} \nu(M)/A(M)$$

exists and is finite no matter how M shrinks to x. It is clear that $K(x)$ then depends continuously on x. If, in addition, $K(x) > 0$ then $K(x)$ is locally bounded away from 0. Because of (5.4) the surface C contains no segment and by (5.1) it is differentiable. Therefore the spherical mapping $x \to u$ maps C topologically on $\nu'(C)$, and $K(x)$ may be considered as a function $K(u)$ on $\nu'(C)$. On Z we can, locally, introduce analytic parameters t_1, t_2 and may then further consider $K(u)$ as a function $K(t) = K(t_1, t_2)$. The inverse $u \to x(u)$ of the spherical mapping induces a representation $x(t) = (x_1(t_1, t_2), x_2(t_1, t_2), x_3(t_1, t_2))$ of C and it is natural to ask whether smoothness of $K(t)$ implies smoothness of $x(t)$. This very difficult question has the following answer:

(5.6) THEOREM. *If $K(t)$ is positive and of class C^m, $m \geqq 2$, then $x(t)$ is at least of class C^{m+1}.*

The first important step in this direction is due to Lewy [2], who proved that a closed surface is analytic if $K(t)$ is analytic.

In (5.6) class C^{m+1} is to be expected instead of C^{m+2} because we know class C^0 for $K(t)$ implies C^1 for $x(t)$ but not C^2 (see last section). Theorem (5.6) with $m \geqq 3$ is due to Pogorelov [4]. The improvement from $m \geqq 3$ to $m \geqq 2$ is obtained by applying the methods of Nirenberg [2]. The, in some respects, most interesting case $m = 1$ is not covered. This is an imperfection which runs through the whole theory in its present stage.

For this reason, and because the complete proof is very long, we will briefly outline the first step which is an existence proof and uses more familiar ideas suggested by the work of S. Bernstein and H. Weyl. We will be more explicit regarding the second step which is Pogorelov's uniqueness proof with the novel feature of admitting arbitrary convex surfaces for competition.

We take the origin z in the interior of the convex set bounded by a complete surface containing C. The supporting function $H(u)$ is the distance from z of the supporting plane to C at the point $x(u)$. As usual we extend $H(u)$ to $x = \lambda u$, $|u| = 1$, $\lambda \geqq 0$ by

$$H(x) = \lambda H(u).$$

Then, see K., p. 62, for C of class C^2 and with $H_{ik} = \partial^2 H(u)/\partial x_i \partial x_k$

$$(5.7) \quad K^{-1}(u) = \begin{vmatrix} H_{11} & H_{12} \\ H_{12} & H_{22} \end{vmatrix} + \begin{vmatrix} H_{22} & H_{23} \\ H_{23} & H_{33} \end{vmatrix} + \begin{vmatrix} H_{33} & H_{31} \\ H_{31} & H_{11} \end{vmatrix}.$$

By methods similar to those of Section 3, Alexandrov proves in [2] that (5.7) is valid at every u where $H(x)$ possesses a second differential.

Theorem (5.6) is clearly of a local nature. We consider therefore a point $u' \in \nu'(C)$ and a small cap δ of Z with center u' such that $\bar{\delta} \subset \nu'(C)$. If we place the x_3-axis in the direction of u' then $t_1 = u_1/u_3$, $t_2 = u_2/u_3$ may serve as analytic coördinates on Z covering a neighborhood of δ. In terms of these, (5.7) becomes, with $h(t_1, t_2) = H(t_1, t_2, 1)$:

$$(5.8) \qquad K^{-1}(t) = (h_{11}h_{22} - h_{12}^2)(1 + t_1^2 + t_2^2)^2,$$

where $h_{ik} = \partial^2 h/\partial t_i \partial t_k$. The cap δ corresponds to an open disk D in the t-plane with the origin as center whose boundary we call B.

If $H(x)$ is differentiable at u then $\partial H/\partial x_i = x_i(u)$, see K., p. 26, so that C will be of class C^{m+1} if $H(x)$ or $h(t)$ is of class C^{m+2}. The method of Nirenberg uses finer distinctions than just the classes C^r. The function $f(y_1, \ldots, y_n)$ is said to satisfy a Hölder condition with exponent α and constant C if

$$|f(y) - f(y')| \leqq C |y - y'|^\alpha.$$

The function $f(y)$ is of class $C^{r+\alpha}$, when r is a positive integer and $0 < \alpha < 1$, if its partial derivatives of order r exist and satisfy Hölder conditions with the exponent α.

The principal lemma of Nirenberg [2] is the following:

If $h(t)$ satisfies (5.8) and is of class C^2 and if $K(t)$ is of class $C^{m+\alpha}$, $m \geqq 2$, $0 < \alpha < 1$, then $h(t)$ is of class $C^{m+2+\alpha}$.

Using this lemma and otherwise the method of Pogorelov the proof proceeds as follows: Assume $K(t)$ to be of class $C^{m+\alpha}$, $m \geqq 2$, $0 < \alpha < 1$. We construct a sequence of analytic functions $K_n(t)$ which uniformly approximate $K(t)$ with its derivatives of order $< m$ in a neighborhood of δ and such that the mth derivatives of $K_n(t)$ satisfy uniform Hölder conditions (i.e., with the same constant) with the exponent α. We also approximate $H(u)$ in D uniformly by supporting functions $H_n(u)$ of analytic convex surfaces, see K., p. 36, or $h(t)$ by $h_n(t)$. For each n a known theorem of S. Bernstein provides an analytic convex surface F_n^δ with supporting function $\bar{h}_n(t)$ and curvature $K_n(t)$ such that $\bar{h}_n(t) = h(t)$ on B.

A priori estimates for the second derivatives of $h_n(t)$ show that these satisfy in every closed subdomain of D a uniform Hölder condition with some exponent β, $0 < \beta < 1$. Passing to a suitable subsequence $\{i\}$ of $\{n\}$ we obtain a sequence $\bar{h}_i(t)$ which tends with its first and second derivatives to a function $\bar{h}(t)$ whose second derivatives satisfy the same Hölder conditions, moreover $\bar{h}(t) = h(t)$ on B. These a priori estimates make the first step of the proof long and they require, unfortunately, $m \geqq 2$.

By Nirenberg's lemma $h(t)$ is of class $C^{m+2+\beta}$. The case where $K(t)$ is of class C^m is obtained from this case by a limit process.

Thus we have a convex surface F_0^δ with supporting function $\bar{h}(t)$ of class C^4 (at least) with the same curvature $K(t)$ as the given surface C (in the part corresponding to D) and such that $\bar{h}(t) = h(t)$ on B. We want to show that $\bar{h}(t) = h(t)$ in D. We observe that $\bar{h}(t)$ and $h(t)$ are convex functions of t and consider the convex surfaces

$$F: \quad w = h(t_1, t_2) \qquad \bar{F}: \quad w = \bar{h}(t_1, t_2)$$

in a space with rectangular coordinates t_1, t_2, w. With $w_{ik} = \partial^2 w / \partial t_i \, \partial t_k$ we know that $\bar{h}(t)$ satisfies everywhere in D, and $h(t)$ at every point of D where it has a second differential, the relation

$$w_{11} w_{22} - w_{12}^2 = K^{-1}(t)(1 + t_1^2 + t_2^2)^{-2}.$$

Assume there are $t \in D$ such that $g(t) = h(t) - \bar{h}(t) < 0$. (The reasoning for points with $g(t) > 0$ is strictly analogous.)

Pogorelov uses an *auxiliary function* $k(t)$, constructed previously by him in [1] for a similar purpose, with the following properties: $k(t)$ is convex and the surface $G : w = k(t)$ intersects $w = 0$ in a closed curve containing D in its interior.

If $w = k'(t)$ is the equation of a supporting plane of G at $(t', k'(t'))$ then with $|t - t'|^2 = (t_1 - t'_1)^2 + (t_2 - t'_2)^2$

$$\liminf_{t \to t'} (k(t) - k'(t)) |t - t'|^{-2} > 0.$$

If t' is a point where $g(t)$ (or $h(t)$) does not have a second differential, then

$$\limsup_{t \to t'} (k(t) - k'(t)) |t - t'|^{-2} = \infty.$$

Unfortunately, the construction of $k(t)$ which is based on (4.12) is much too intricate to be reproduced here.

For a sufficiently small $\epsilon > 0$

$$k_2(t) = g(t) - \epsilon k(t) = h(t) - \bar{h}(t) - \epsilon k(t)$$

has a negative minimum a. The point in introducing $k(t)$ is that for $k_\epsilon(t') = a$ the function $h(t)$ must have at t' a second differential. For the minimum property of t' implies the existence of supporting planes $w=k'(t)$, $w=h'(t)$ of G at $(t', k(t'))$ and of F at $b=(t', h(t'))$ such that these planes and the tangent plane $w = \bar{h}'(t)$ of $\bar{F}$ at $(t', h'(t'))$ satisfy

$$\epsilon k'(t) = h'(t) - \bar{h}'(t) - a, \text{ hence}$$

$$\begin{aligned} (5.9) \quad & \epsilon (k(t) - k'(t)) \\ & = a - k_\epsilon(t) + h(t) - h'(t) - (\bar{h}(t) - \bar{h}'(t)) < h(t) - h'(t). \end{aligned}$$

Therefore, if $h(t)$ did not possess a second differential at t' the properties of $k(t)$ would yield

$$\limsup_{t \to t'} (h(t) - h'(t)) |t - t'|^{-2} = \infty,$$

$$\liminf_{t \to t'} (h(t) - h'(t)) |t - t'|^{-2} > 0.$$

The first of these relations means that the lower indicatrix of F at b contains b on the boundary and the second that the upper indicatrix is bounded. It is plausible and not hard to see that then $\nu(M)/A(M)$ becomes arbitrarily large for suitable Borel sets M containing b so that the specific curvature of F is not bounded. On the other hand, when M' lies in the part of C corresponding to $\bar{D}$, then $\nu(M')/A(M')$ is bounded; it is readily verified that this implies boundedness of $\nu(M)/A(M)$ for $M \subset F$.

Thus $h(t)$ possesses at t' a second differential, so that

$$\bar{h}_{11}\bar{h}_{22} - \bar{h}_{12}^2 = h_{11}h_{22} - h_{12}^2 > 0 \text{ at } t'.$$

Of course, F possesses at t' a tangent plane. We deduce from (5.9) that

$$h(t) - \bar{h}(t) - (h'(t) - \bar{h}'(t)) > \epsilon\,(k(t) - k'(t)).$$

Hence the quadratic form with coefficients $h_{ij}(t') - \bar{h}_{ij}(t')$ is positive definite. But this is impossible because the forms with coefficients $h_{ij}(t')$ and $\bar{h}_{ij}(t')$ are positive definite and have equal discriminants.

CHAPTER II

The Brunn-Minkowski Theory and Its Applications

6. *Mixed volumes*

Pogorelov's theorem (5.6) asserts the smoothness of an already existing, not necessarily complete, convex surface with a smooth positive curvature. Minkowski's problem deals with the existence and uniqueness of a general closed convex hypersurface, whose curvature is given as a function of the unit normal, i.e., on all of Z. This problem, which includes providing one adequate definition or substitute for curvature of general convex surfaces, will be solved in Section 8. In this section we list the necessary tools from the theory of convex bodies found in Bonnesen and Fenchel's book, K. In the next we principally prove results which were discovered after K. appeared.

In a given rectangular coordinate system $x, \ldots, x_n$ in E^n we put

$$x \cdot y = \sum_{i=1}^{n} x_i y_i .$$

Using oriented hyperplanes proves convenient: the positive side of the hyperplane

$$P: \qquad x \cdot v = a, \ v \neq 0$$

is the side $x \cdot v > a$, i.e., the side into which v points if laid off from a point of P. $-x \cdot v = x \cdot (-v) = -a$ represents the other orientation of the same hyperplane. a is positive when the origin z lies on the negative side of P.

A *convex body* K is a bounded closed convex set not necessarily with interior points. For a given vector $v \neq 0$ there is exactly one oriented supporting plane P_v of K with normal v. The equation of P_v may be written in the form

$$P_v: \qquad x \cdot v = H(v)$$

An alternate definition of $H(v)$ is

$$H(v) = \max_{x \in K} x \cdot v,$$

which shows that $H(\lambda v) = \lambda H(v)$ for $\lambda > 0$ and suggests the agreement $H(0) = 0$, see K., pp. 23, 24. The function $H(v)$ is the so-called *supporting function* of K; it is positive for $v \neq 0$ when z is an interior point of K. If u is a unit vector then $H(u)$ is the signed distance of P_u from z, so that our present terminology agrees with that of the last section.

The relation $H(u) = -H(-u)$ is evidently equivalent to the statement that the non-oriented hyperplanes carrying P_u and P_{-u} coincide, hence K lies in that hyperplane. It is useful to notice that $a \cdot v$ is the supporting function of the point a and that $|v|$ is the supporting function of the unit sphere or ball $|x| \leqq 1$.

The principal property of $H(v)$ is its convexity, which owing to the homogeneity of $H(v)$ can be expressed in the simple form:

$$H(v + w) \leqq H(v) + H(w) \tag{6.1}$$

and follows immediately from $H(v) = \max x \cdot v$.

A function $H(v)$ defined in the whole space and satisfying the condition $H(\lambda v) = \lambda H(v)$ for $\lambda \geqq 0$ and (6.1) is the supporting function of exactly one convex body, K., Section 17.

For any set M in E^n we denote for real λ by λM the set consisting of the points λx, $x \in M$, and for two sets M, N by $M + N$ the sets of points $x + y$ with $x \in M$ and $y \in N$. In particular, if N is the point a, then $M + a$ originates from N by the translation $x' = x + a$.

If K_i are convex bodies in E^n and $\lambda_i \geqq 0$, $i = 1, \ldots, r$, then

$$K = \lambda_1 K_1 + \ldots + \lambda_r K_r$$

is a convex body, K., Section 20. The preceding remark shows that K depends on the position of z; changing z induces a translation of K. The properties in which we will be interested, are invariant under translations.

If $H_i(v)$ is the supporting function of K_i then $\Sigma \lambda_i H_i(v)$ is the supporting function of K. Therefore $H_1(v) + a \cdot v$ is the *supporting function of* $K_1 + a$ and $H_1(v) + \epsilon\,|v|$ is the supporting function of the closure of the ϵ-neighborhood of K_1. Consequently a sequence of convex bodies K_i, $i = 1, 2, \ldots$, with supporting functions $H_i(v)$ tends to the convex body K with supporting function $H(v)$ in the sense of Hausdorff's closed limit, if and only if $H_i(u)$ tends, uniformly on Z, to $H(u)$. Any bounded sequence $\{K_i\}$ (i.e., $K_i \subset U(z, R)$ for a suitable R) contains a converging subsequence, K., p. 34.

We come now to the *basic ideas of Minkowski*, K., Section 29. The n-dimensional measure or volume of K is denoted by $V(K)$. For variable $\lambda_i \geqq 0$ the volume of $K = \Sigma_{i=1}^{r} \lambda_i K_i$ is a form

$$V(K) = \sum_{i_1=1}^{r} \sum_{i_2=1}^{r} \cdots \sum_{i_n=1}^{r} V_{i_1 \ldots i_n} \lambda_{i_1} \cdots \lambda_{i_n}$$

of degree n in the λ_i, where the coefficients $V_{i_1 \ldots i_n}$ are uniquely determined by requiring that they are symmetric in their subscripts. Then $V_{i_1 \ldots i_n}$ *depends only on the bodies* $K_{i_1}, \ldots, K_{i_n}$ *and not on the remaining bodies* K_j, *hence we may write* $V(K_{i_1}, \ldots, K_{i_n})$ for $V_{i_1 \ldots i_n}$ *and call it the mixed volume of* $K_{i_1}, \ldots, K_{i_n}$. Although i runs originally from 1 to r we see that studying mixed volumes in E^n amounts to studying $V(K_1, \ldots, K_n)$ where the K_i are not necessarily distinct. In fact, the most important case is that with only two distinct K_i, so that it pays to introduce the notation:

$$V_m(K_1, K_2) = V(\underbrace{K_1, \ldots, K_1}_{n-m}, \underbrace{K_2, \ldots, K_2}_{m}) = V_{n-m}(K_2, K_1).$$

Then

$$V(\lambda_1 K_1 + \lambda_2 K_2) = \sum_{m=0}^{n} \binom{n}{m} \lambda_1^{\,n-m} \lambda_2^{\,m} V_m(K_1, K_2)$$

and

$$n\, V_1(K_1, K_2) = \lim_{h \to 0+} h^{-1}[V(K_1 + h\, K_2) - V(K_1)]. \tag{6.2}$$

Since translating each K_i in an arbitrary way induces a translation of $K = \Sigma \lambda_i K_i$, *the mixed volumes are invariant under such translations, but not under other independent motions of the* K_i.

For example, the sum of the two proper parallel segments is a segment parallel to these, whereas the sum of two non-parallel segments is a parallelogram.

We notice that

(6.3) $$V(K_1, \ldots, K_{n-1}, a) = 0 \text{ for a point } a,$$

because evidently

$$V(\lambda_1 K_1 + \ldots + \lambda_{n-1} K_{n-1}) = V(\lambda_1 K_1 + \ldots + \lambda_{n-1} K_{n-1} + \lambda_n a).$$

Since the volume of a convex body depends continuously on the body the same holds for the mixed volumes. $V(\Sigma \lambda_i K) = (\Sigma \lambda_i)^n \, V(K)$ entails

$$V(K, \ldots, K) = V(K), \text{ or } V_0(K_1, K_2) = V(K_1), V_n(K_1, K_2) = V(K_2).$$

Not so simple is the *monotoneity*

$$V(K_1, K_2, \ldots, K_n) \leqq V(K'_1, K'_2, \ldots, K'_n) \text{ for } K_i \subset K'_i, i = 1, \ldots, n.$$

Together with (6.3) it implies that the mixed volumes are non-negative; we need only to replace K_n by a point $a \in K_n$. Clearly, $V(K_1, \ldots, K_n) > 0$ if and only if we can find proper segments $T_i \subset K_i$ which are not all parallel to the same hyperplane, see K., p. 41.

Consider a non-degenerate convex polyhedron P, i.e., the convex closure of a finite number of points which do not lie in a hyperplane. Denote by $p^1, \ldots, p^N$ the faces of P, by $v(p^i)$ the $(n-1)$-dimensional volume or area of p^i, by w^i the unit normal to the supporting plane of P containing p^i. If $H_P(v)$ is the supporting function of P, then P is the intersection of the half spaces $x \cdot w^i \leqq H_P(w^i)$ and

$$V(P) = n^{-1} \sum_{i=1}^{N} v(p^i) H_P(w^i).$$

Very important (K., Section 29) is the *generalization of this relation to mixed volumes.* If K^* is any convex body and $H^*(v)$ its supporting function then

(6.4) $$V_1(P, K^*) = V_{n-1}(K^*, P) = n^{-1} \sum_{i=1}^{N} H^*(w^i) v(p^i).$$

This relation generalizes further if applied to the linear combination $P = \lambda_1 P_1 + \ldots + \lambda_{n-1} P_{n-1}$, $\lambda_i > 0$, of non-degenerate polyhedra P_i, which is again a polyhedron. We use for P the same notation as before, denote by $p_j{}^i$ the intersection of P_j with its supporting plane with normal w^i. Then, K., p. 31,

$$p^i = \sum_{j=1}^{n-1} \lambda_j p_j{}^i.$$

Hence the $(n-1)$-dimensional mixed volume $v(p^i{}_1, \ldots, p_{n-1}{}^i)$ is defined. Applying (6.4) and comparing the coefficients of $\lambda_1, \ldots, \lambda_{n-1}$ yields then

$$(6.5)\quad V(K^*, P_1, \ldots, P_{n-1}) = n^{-1} \sum_{i=1}^{N} H^*(w^i) v(p_1{}^i, \ldots, p_{n-1}{}^i).$$

If K is an arbitrary convex body with interior points, ∂K its boundary, dS the area element of ∂K, then a limit process leads from (6.4) to

$$(6.6)\qquad V_1(K_1, K^*) = V_{n-1}(K^*, K) = n^{-1} \int_{\partial K} H^*(w) dS.$$

If K is strictly convex, then ∂K is represented in terms of the supporting function $H(v)$ of K by

$$(6.7)\qquad x_i = \frac{\partial H(v)}{\partial v_i}, \quad i = 1, \ldots, n,$$

where v need not be a unit vector because $\partial H/\partial v_i$ is homogeneous of degree 0. We used (6.7), see K., p. 26, already in the last section. We assume now in addition that $H(v)$ is for $v \neq 0$ of class C^2 and put

$$H_{ij}(v) = \partial H(v)/\partial v_i\, \partial v_j;$$

the H_{ij} are homogeneous of order -1. Then (6.7) and the Euler relations for homogeneous functions yield, K., Section 37, that

$$(6.8)\qquad V_1(K, K^*) = n^{-1} \int_Z H^*(w) D_{n-1}(H, w) d\sigma,$$

where $d\sigma$ is the area element of the unit sphere Z at w, and $D_m(H, w)$ or simply $D_m(H)$ denotes generally the sum of all principal minors of the order m of the matrix $H_{ij}(w)$.

The same argument which leads from (6.4) to (6.5) leads from (6.8) to

(6.9) $V(K_1, \ldots, K_n) = n^{-1}\int_Z H^{(n)}(w)D(H^{(1)}, \ldots, H^{(n-1)}, w)\, d\sigma,$

if K_i, $i = 1, \ldots, n-1$, are non-degenerate strictly convex bodies with supporting functions $H^{(i)}$ of class C^2 and $H^{(n)}$ is the supporting function of K_n. Here $D(H^{(1)}, \ldots, H^{(n-1)}, w)$ is the factor of $\lambda_1 \ldots \lambda_{n-1}$ in $D_{n-1}(\lambda_1 H^{(1)} + \ldots + \lambda_{n-1}H^{(n-1)})$ divided by $(n-1)!$. We conclude from (6.3) that (6.9) vanishes when K_n is an arbitrary point a, or $H^{(n)}(w) = a \cdot w$, so that

(6.10) $\int_Z w_i D(H^{(1)}, \ldots, H^{(n-1)}, w) d\sigma = 0, \; i = 1, \ldots, n.$

If $R_1, \ldots, R_{n-1}$ are the *principal radii of curvature* of ∂K at $x(w)$ then $dS = R_1 \ldots R_{n-1} d\sigma$, so that we expect the relation $D_{n-1}(H) = R_1 \ldots R_{n-1}$ which we used for $n = 3$ in the last section. This follows from the formula of Olindes-Rodrigues $dx = Rdw$ for the principal directions, which by (6.7) may be written as

$$\sum_{k=1}^{n} H_{ik}(w) dw_k = R dw_i.$$

Therefore the R_i satisfy the equation

$$\begin{vmatrix} H_{11} - R & H_{12} & \ldots & H_{1n} \\ H_{21} & H_{22} - R & \ldots & H_{2n} \\ H_{n1} & H_{n2} & \ldots & H_{nn} - R \end{vmatrix} = 0$$

From homogeneity $\Sigma_k H_{ik} w_k = 0$, hence the determinant $| H_{ik} |$ vanishes, and $R = 0$ is one root. Factoring this root out leaves

$$R^{n-1} - D_1(H) R^{n-2} + D_2(H) R^{n-3} + \ldots + (-1)^{n-1} D_{n-1}(H) = 0$$

which shows that

(6.11) $D_j(H) = \{R_1 \ldots R_j\},$

where $\{R_1 \ldots R_j\}$ is the *elementary symmetric function of the* R_i of which $R_1 \ldots R_j$ is one summand.

The *significance of the mixed volumes* lies in their flexibility.

Substituting: suitable bodies $K_2, \ldots, K_n$ in $V(K_1, \ldots, K_n)$ gives various geometric quantities attached to K_1. The principal ones for us are the $V_m(K, U)$ where U is the unit ball $|x| \leqq 1$, of which the simplest, $nV_1(K, U)$, is by (6.6) the area of ∂K because $H^*(w) = |w| \equiv 1$ for U. This may also be seen as follows: $K + hU$ is the closed h-neighborhood of K, hence $(K + hU) - K$ the shell about K with thickness h, and from (6.2)

$$\lim_{h \to 0+} h^{-1}[V(K + hU) - V(K)] = n\, V_1(K, U).$$

To find a geometric interpretation for all $V_m(K, U)$ we use the relation, see K., pp. 62, 63,

$$(6.12)\quad \{R_1 \ldots R_m\} = \binom{n-1}{m} D(\underbrace{|w|, \ldots, |w|}_{n-m-1}, \underbrace{H, \ldots, H}_{m}) = D_m(H),$$

where H is again the supporting function of K. Therefore (6.9) with K^n as U or as K yields for $m = 1, 2, \ldots, n-2$, that

$$(6.13)\quad V_m(K, U) = n^{-1}\binom{n-1}{n-m}^{-1} \int_Z \{R_1 \ldots R_{n-m}\} d\sigma$$

$$= n^{-1}\binom{n-1}{n-m-1}^{-1} \int_Z H\{R_1 \ldots R_{n-m-1}\} d\sigma,$$

and for $m = n-1$ that

$$(6.14)\quad V_{n-1}(K, U)$$

$$= n^{-1}(n-1)^{-1} \int_Z (R_1 + \ldots + R_{n-1}) d\sigma = 2^{-1} n^{-1} \int_Z B(w) d\sigma,$$

where $B(w) = H(w) + H(-w)$ is the distance of the supporting planes of K with normals w and $-w$, the so-called width of K in the direction $\pm w$.

There are analogous expressions for $V_m(K, U)$ if $m < n - 1$: The width $B(u)$ is the projection of K on the line $x = tu$ by hyperplanes normal to this line. If V is an $(n - m)$-flat through z we define $B(V)$ as the $(n - m)$-dimensional volume of the projection of K on V by m-flats normal to V. Then $V_m(K, U)$ is, except for a factor depending only on n, the average of $B(V)$ over all $(n - m)$-flats V through z, see K., pp. 46, 50. The case $m = 1$

gives Cauchy's formula for the area of ∂K as the average of the projections of K on the hyperplanes through z, K., p. 48.

We notice the following consequence of (6.10) and (6.12):

(6.15) $\int_Z w_i\{R_1 \ldots R_m\} d\sigma = 0, \ m = 1, \ldots, n-1; \ i = 1, \ldots, n.$

7. *The general Brunn-Minkowski theorem*

There are some very important inequalities between the mixed volumes. The older ones follow from the

(7.1) Brunn-Minkowski Theorem. *If K and L are convex bodies in E^n then*

$$g(\varrho) = V^{1/n}((1-\varrho)K + \varrho L)$$

is for $0 \leqq \varrho \leqq 1$ a concave function of ϱ, which is linear if and only if K and L are homothetic or lie in parallel hyperplanes.

For an, in view of the many non-trivial implications, comparatively simple proof, see K., Section 48. The tangent of the graph of

$$g(\varrho) = \left[\sum (1-\varrho)^{n-m} \varrho^m \binom{m}{n} V_m(K, L)\right]^{1/n}$$

at $\varrho = 0$ lies therefore over the curve and passes through $(1, g(1))$ only when $g(\varrho)$ is linear. This yields *Minkowski's inequalities*:

(7.2) $V_1{}^n(K, L) \geqq V^{n-1}(K)V(L), \quad V_{n-1}^n(K, L) \geqq V(K)V^{n-1}(L)$

If K and L do not lie in parallel hyperplanes the equality sign holds only when K and L are homothetic.

The fact that the second derivative at 0 of $g(\varrho)$ is non-positive yields the further, so-called *quadratic, inequalities of Minkowski*:

(7.3) $V_1{}^2(K, L) \geqq V(K)V_2(K, L), \ V_{n-1}^2(K, L) \geqq V_{n-2}(K, L)V(L).$

The equality holds in the first of these inequalities (we assume that K and L do not lie in parallel hyperplanes) not only when K and L are homothetic but also when K is homothetic to a Kappenkörper of L, see K., pp. 17, 92. Minkowski conjectured, and Bol [1] proved, that this is the only case where equality enters.

These inequalities solve many extremal and uniqueness

problems, see K., and are doubtless not as well known as they ought to be. Here we will use them only to prove uniqueness of the solution of Minkowski's problem.

Conjectures on extensions of these results to other mixed volumes were current for some time, see K., pp. 92, 93. They were proved independently by Fenchel [1, 2] and Alexandrov [1]. Both establish first the following generalization of (7.3):

$$(7.4)\quad V^2(C_1,\dots,C_{n-2},K,L) \geqq V(C_1,\dots,C_{n-2},K,K)V(C_1,\dots,C_{n-2},L,L),$$

where $C_1, \dots, C_{n-2}, K, L$ *are, of course, convex bodies in* E^n.

It is quite easy to see that (7.4) implies generalizations of (7.1) and (7.2): We put

$$V_{m,k}(C,K,L)$$
$$= V(C_1,\dots,C_{n-m},\underbrace{K,\dots,K}_{m-k},\underbrace{L,\dots,L}_{k}),\ 0 \leqq m \leqq n,\ 0 \leqq k \leqq m$$

so that (7.4) becomes

$$V^2_{2,1}(C,K,L) \geqq V_{2,0}(C,K,L)\, V_{2,2}(C,K,L)$$

and show first that (7.4) contains the

(7.5) General Brunn-Minkowski Theorem. *If* $K_\rho = (1-\varrho)K + \varrho L$, *then*

$$f(\varrho) = V^{1/m}_{m,0}(C,K_\rho,L) = V^{1/m}(C_1,\dots,C_{n-m},K_\rho,\dots,K_\rho),\quad m \geqq 2,$$

is for $0 \leqq \varrho \leqq 1$ *a concave function of* ϱ.

We show that $f''(\varrho) \leqq 0$. It suffices to prove this for $\varrho = 0$. For if $0 < \varrho_1 < 1$, then K_ρ can for $\varrho_1 \leqq \varrho \leqq 1$ with $\varrho' = (\varrho - \varrho_1)(1-\varrho_1)^{-1}$ be written as

$$K_\rho = \bar{K}_{\rho'} = (1-\varrho')K_{\rho_1} + \varrho' L$$

and if $\bar{f}(\varrho') = V^{1/m}_{m,0}(C,\bar{K}_{\rho'},L)$ then

$$\bar{f}''(0) = (1-\varrho_1)^2 f''(\varrho_1),$$

so that $f''(\varrho_1) \leqq 0$ if $\bar{f}''(0) \leqq 0$. Now obviously, see also K., p. 40,

$$(7.6)\quad V_{m,0}(C,K_\rho,L) = \sum_{k=0}^{m} (1-\varrho)^{m-k}\varrho^k \binom{m}{k} V_{m,k}(C,K,L)$$

and therefore

$$f''(0) = (m-1)\, V_{m,0}^{1/m-2}(V_{m,0}V_{m,2} - V_{m,1}^2)$$

so that $f''(0) \leqq 0$ and hence (7.5) follows from (7.4) with $C_{n-m+1} = \ldots = C_{n-2} = K$. The tacit assumption $V_{m,0} > 0$ is readily removed by a limit process. The relation $f'(0) \geqq f(1) - f(0)$ gives

$V_{m,1}^m \geqq V_{m,0}^{m-1} V_{m,m}$ with equality only when $f(\varrho)$ is linear.

If in (7.4) we put $m - k - 1$ of the C_i equal to K and $k - 1$ equal to L we find

$$(7.7) \qquad V_{m,k}^2(C, K, L) \geqq V_{m,k-1}(C, K, L)\, V_{m,k+1}(C, K, L)$$

and this yields, exactly as before, the following generalization of the preceding relation:

$$(7.8) \qquad V_{m,k}^m(C, K, L) \geqq V_{m,0}^{m-k}(C, K, L)\, V_{m,m}^k(C, K, L)$$

with equality only when the corresponding function $f(\varrho)$ is linear.

Alexandrov noticed in A. [1, II] that these inequalities lead to an easily remembered more *general inequality*:

$$(7.9) \quad V^m(C_1, \ldots, C_n) \geqq \prod_{k=0}^{m-1} V(C_1, \ldots, C_{n-m}, C_{n-k}, \ldots, C_{n-k})$$

or

$$V_{0,0}^m(C, C_{n-1}, C_n) = V_{2,1}^m(C, C_{n-1}, C_n) \geqq$$

$$\geqq \prod_{k=0}^{m-1} V_{m,0}(C, C_{n-k}, C_{n-m}) = \prod_{k=0}^{m-1} V_{m+1,1}(C, C_{n-k}, C_{n-m}).$$

For $m = 2$ this becomes (7.4) with $K = C_n$, $L = C_{n-1}$. Assume therefore that (7.9) holds for some $m \geqq 2$. Then (7.8) yields

$$V_{0,0}^{m(m+1)} \geqq \prod_{k=0}^{m-1} V_{m+1,1}^{m+1}(C, C_{n-k}, C_{n-m})$$

$$\geqq \prod_{k=0}^{m-1} V_{m+1,0}^m(C, C_{n-k}, C_{n-m})\, V_{m+1,m+1}(C, C_{n-k}, C_{n-m})$$

$$= \prod_{k=0}^{m} V_{m+1,0}^m(C, C_{n-k}, C_{n-m+1})$$

which is (7.9) for $m + 1$ raised to the power m.

For $m = n$ (7.9) becomes

$$V^n(C_1, \ldots, C_n) \geqq V(C_1)\, V(C_2) \ldots V(C_n).$$

The equality sign in (7.9) for any $m > 2$ requires equality in (7.8) hence linearity of $f(\varrho)$. Calling a convex body K or its boundary ∂K *regular* if K contains interior points and ∂K is a surface of class C^2 with positive curvature, the following is known regarding linearity of $f(\varrho)$, see A. [1, IV].

(7.10) *If K and L are convex bodies of dimension at least m and if $C_1, \ldots, C_{n-m}$ are regular, then*

$$f(\varrho) = V^{1/m}(C_1, \ldots, C_{n-m}, K_\rho, \ldots, K_\rho), K_\rho = (1-\varrho)K + \varrho L,$$

is linear if, and only if, K and L are homothetic.

We will prove (7.10) only under the assumption that K and L are also regular.

Three proofs of the basic inequality (7.4) are available, one by Fenchel [1] and two by Alexandrov [1, II, IV]. Fenchel's proof is very sketchy, a detailed version has never appeared and it is not quite clear what it would involve. The simplest one, at present, seems to be the second proof of Alexandrov [1, IV] which generalizes Hilbert's approach to (7.3), see Hilbert [1], Chapter 19 or K., pp. 102–104. Even this proof is too long to be reproduced here in all details in view of the few applications which we will make of (7.4) or rather (7.7). But we will discuss all the essential steps.

The proof is based on the relation (6.9) and investigates first the properties of $D(H^{(1)}, \ldots, H^{(n-1)}, w)$. Consider r real quadratic forms

$$q_k = \sum_{i,j=1}^{n} a_{ij}^{(k)} x_i x_j, \; k = 1, \ldots, r, \; a_{ij}^{(k)} = a_{ji}^{(k)},$$

in n variables. For any real $\lambda_1, \ldots, \lambda_r$

$$q = \lambda_1 q_1 + \ldots + \lambda_r q_r = \sum_{i,j=1}^{n} a_{ij} x_i x_j, \; a_{ij} = \sum_{k=1}^{r} \lambda_k a_{ij}^{(k)}$$

is also a quadratic form. Its *discriminant* can be written in the form:

$$D(q) = \sum_{i_1=1}^{r} \cdots \sum_{i_n=1}^{r} \lambda_{i_1} \ldots \lambda_{i_n} D(q_{i_1}, \ldots, q_{i_n}), \tag{7.11}$$

where $D(q_{i_1}, \ldots, q_{i_n})$ is independent of the order of the q_{i_k}, and depends only on $q_{i_1}, \ldots, q_{i_n}$ (this is seen by putting λ_j with $j \neq i_k$ equal to zero). $D(q_i, \ldots, q_i) = D(q_i)$ because $\Sigma \lambda_{i_1} \ldots \lambda_{i_n} = (\lambda_1 + \ldots + \lambda_r)^n$. In analogy to the mixed volumes, $D(q_{i_1}, \ldots, q_{i_n})$ is called the *mixed discriminant* of $q_{i_1}, \ldots, q_{i_n}$. Studying its properties amounts to investigating $D(q_1, \ldots, q_n)$ where the q_i are not necessarily distinct.

If the variables $x_1, \ldots, x_n$ undergo a non-degenerate homogeneous linear transformation—for brevity we call such transformations *central affinities*—with determinant Δ then $D(q_1, \ldots, q_n)$ is multiplied by Δ^2, hence does not change if $\Delta = \pm 1$.

Obviously $D(q_1, \ldots, q_n)$ is linear in the coefficients of each form:

$$(7.12) \qquad D(q_1, \ldots, q_n) = \sum_{i,j=1}^{n} D(q_1, \ldots, q_{n-1})_{ij} \, a_{ij}^{(n)}$$

and a simple calculation with determinants shows that $D(q_1, \ldots, q_{n-1})_{ii}$ is the mixed discriminant of the forms in $n - 1$ variables obtained from $q_1, \ldots, q_{n-1}$ by putting $x_i = 0$. Most important is

(7.13) *If $q_1, \ldots, q_n$ are positive definite, then $D(q_1, \ldots, q_n) > 0$.*

We prove this by induction: the assertion is trivial for $n = 1$. Assume (7.13) to hold for $n - 1$ variables. A unimodular, central affinity will bring q_n into the form $\Sigma_{i=1}^{n} b_i x_i^2$ which we call canonical, with $b_i > 0$ because q_n is positive definite. If this affinity takes q_i for $i < n$ into q'_i then by (7.12)

$$D(q_1, \ldots, q_n) = D(q'_1, \ldots, q'_n) = \sum D(q'_1, \ldots, q'_{n-1})_{ii} \, b_i,$$

and $D(q'_1, \ldots, q'_{n-1})_{ii} > 0$ because the forms originating from q'_i by putting $x_n = 0$ are still positive definite, and by the inductive hypothesis.

Rather simple operations with matrices and determinants yield:

(7.14) *A unimodular central affinity with matrix M performed on the quadratic form*

$$(7.15) \qquad \sum_{i,k=1}^{n} D(q_1, \ldots, q_{n-1})_{ik} \, x_i x_k$$

has the same effect as the central affinity with matrix ${}^tM^{-1}$ performed *on the forms* $q_1, \ldots, q_{n-1}$ *where* ${}^tM$ *is the transpose of* M. *In particular,* $D(q'_1, \ldots, q'_{n-1})_{ik} = D(q_1, \ldots, q_{n-1})_{ik}$ *if* $M = {}^tM^{-1}$.

It follows that $q_1, \ldots, q_{n-1}$ can be transformed such that $D(q'_1, \ldots, q'_{n-1})_{ik} = 0$ for $i \neq k$. If the forms $q_1, \ldots, q_{n-1}$ are positive definite, then $D(q'_1, \ldots, q'_{n-1}) > 0$ by (7.13) hence $\Sigma_i D(q'_1, \ldots, q'_{n-1})_{ii} x_i^2$ is positive definite and we conclude from (7.14):

(7.16) *The form* (7.15) *is positive definite if* $q_1, \ldots, q_{n-1}$ *are positive definite.*

We now come to the decisive fact on mixed discriminants, namely the analogue to (7.4):

(7.17) THEOREM. *If the forms* $q_1, \ldots, q_{n-1}$ *are positive definite and* $Q = \Sigma b_{ik} x_i x_k$, $b_{ik} = b_{ki}$, *is any form then*

$$D^2(q_1, \ldots, q_{n-1}, Q) \geqq D(q_1, \ldots, q_{n-1}, q_{n-1})\, D(q_1, \ldots, q_{n-2}, Q, Q);$$

the equality sign holds only when $Q = \lambda q_{n-1}$.

Since $D(q_1, \ldots, q_{n-1}, q_{n-1}) > 0$ by (7.13), we deduce from (7.17):

(7.18) $D(q_1, \ldots, q_{n-1}, Q) = 0$ implies $D(q_1, \ldots, q_{n-2}, Q, Q) \leqq 0$, with equality only for $Q \equiv 0$.

Actually (7.18) is equivalent to (7.17). To see this and for the proof of (7.18) we introduce the abbreviations

$$D_p(q, Q) = D(q_1, \ldots, q_{n-p}, \underbrace{Q, \ldots, Q}_{p}),$$

$$D_p(q, Q)_{ik} = D(q, \ldots, q_{n-p-1}, \underbrace{Q, \ldots, Q}_{p})_{ik}$$

Define λ by $D_1(q, Q) = \lambda D_1(q, q_{n-1})$. Then $D_1(q, Q - \lambda q_{n-1}) = 0$, hence (7.18) yields $D_2(q, Q - \lambda q_{n-1}) \leqq 0$. Expanding this and substituting the value for λ leads to

$$D_2(q, Q) - D_1^2(q, Q)\, D_1^{-1}(q, q_{n-1}) \leqq 0,$$

hence to (7.17) with the condition for the equality sign.

The proof of (7.18) proceeds by *induction with respect to* n and is quite long. The case $n = 2$ is very simple and is left to the reader. Assume that (7.18) holds for $n - 1 \geqq 2$ variables. Using (7.12) we see that $D_2(q, Q)$ is a quadratic form G in the b_{ij}

$$G = \sum_{i,k=1}^{n} D_1(q, Q)_{ik} b_{ik} = \sum_{i,k,j,l} D(q_1, \ldots, q_{n-2})_{ik,jl} b_{ik} b_{jl},$$

where $D(q_1, \ldots, q_{n-2})_{ii,ii} = 0$ and $D(q_1, \ldots, q_{n-2})_{ii,jj} > 0$ for $i \neq j$, because it is the mixed discriminant of the forms obtained from $q_1, \ldots, q_{n-2}$ by putting $x_i = 0$, $x_j = 0$.

The next step is a study of the *eigenvalues* of G as a quadratic form in the variables b_{ik}, see Courant-Hilbert [1], Chapter 1.

(7.19) *0 is not an eigenvalue of G.*

For 0 is an eigenvalue if, and only if, the linear equations

$$D_1(q, Q)_{ik} = 0$$

in the b_{ik} have a non-trivial solution. A form Q_0 whose coefficients are solutions of these equations is called an *eigenform* of $D_2(q, Q)$. We have to show that $Q_0 \equiv 0$. A central affinity will take q_{n-2} and Q into the canonical forms $q'_{n-2} = \Sigma a_{ii}^{(n-2)} x_i^2$, $a_{ii}^{(n-2)} > 0$, and Q'_0, and q_i into certain forms q'_i for $i = 1, \ldots, n-3$. If 0 is an eigenvalue of G then it is also one of $G' = D_2(q', Q)$ and Q'_0 will be an eigenform of G'. Therefore

$$D_1(q', Q'_0)_{ii} = 0, \ i = 1, \ldots, n$$

and by the inductive assumption

$$D_2(q', Q'_0) \leqq 0, \ i = 1, \ldots, n, \tag{7.20}$$

with equality for all i only if $Q'_0 \equiv 0$. Since q'_{n-2} has canonical form,

$$D_2(q', Q'_0) = \sum D_2(q', Q'_0)_{ii} \, a_{ii}^{(n-2)}.$$

The left side vanishes because Q'_0 is eigenform of G', hence the right side vanishes, which implies $D_2(q', Q'_0)_{ii} = 0$ for all i, and (7.19) follows.

(7.21) *G has exactly one positive eigenvalue.*

Here, and in similar statements, it is understood that the eigenvalues are counted with their multiplicities, so that (7.21) means that the sum of the multiplicities of all positive eigenvalues s 1.

It is easily verified that $D_2(\Sigma_i x_i^2, Q)$ has exactly one positive eigenvalue. The forms

$$q_i^\rho = (1-\varrho) \sum x_i^2 + \varrho q_i, \quad i = 1, \ldots, n-2, \ 0 \leqq \varrho \leqq 1,$$

are positive definite and the eigenvalues of $D_2(q^\rho, Q)$ vary continuously with ϱ. By (7.19) no eigenvalue vanishes for any ϱ whence (7.21) follows.

We now complete the proof of (7.18). Because the number of positive eigenvalues of a quadratic form does not increase if certain variables are put equal to 0 the form

$$(7.22) \qquad \sum_{i,k} D(q_1, \ldots, q_{n-2})_{ii,kk}\, b_{ii}\, b_{kk}$$

has by (7.21) at most one positive eigenvalue. By a unimodular affinity we put q_{n-1} and Q (as given in (7.18)) into canonical form. It suffices to prove (7.18) in this case because the mixed discriminants do not change under this affinity. Then $D_2(q, Q)$ takes the form (7.22). The form

$$(7.23) \qquad \sum [a_{ii}^{(n-1)}]^{-1} D_1(q, q_{n-1})_{ii}\, b_{ii}^2 = \sum c_i\, b_{ii}^2$$

is positive definite by (7.13) and because $a_{ii}^{(n-1)} > 0$. By a central affinity operating on the b_{ii} we can put both (7.23) and (7.22) into canonical form. This amounts to solving the equations

$$(7.24) \qquad \sum_{k=1}^{n} D(q_1, \ldots, q_{n-2})_{ii,\,kk}\, b_{ii}\, b_{kk} = \lambda c_i\, b_{ii}.$$

Since

$$D_1(q, q_{n-1})_{ii} = \sum_{k=1}^{n} D(q_1, \ldots, q_{n-2})_{ii,\,kk}\, a_{kk}^{(n-1)},$$

the equations (7.24) have for $\lambda = 1$ the solutions $b_{kk} = a_{kk}^{(n-1)}$. We know that (7.22) has at most one positive eigenvalue, consequently $\lambda = 1$ is the only one. The hypothesis $D_1(q, Q) = 0$ of (7.18) may be written as

$$\sum_{i=1}^{n} [a_{ii}^{(n-1)}]^{-1} D_1(q, q_{n-1})\, a_{ii}^{(n-1)}\, b_{ii} = \sum_i c_i\, a_{ii}^{(n-1)}\, b_{ii} = 0,$$

so that the vector $b = (b_{11}, \ldots, b_{nn})$ is in the sense of the metric given by the form (7.23) orthogonal to the vector $a^{(n-1)} = = (a_{11}^{(n-1)}, \ldots, a_{nn}^{(n-1)})$ belonging to the only positive eigenvalue.

The next smaller eigenvalue is the minimum of $G = D_2(q, Q)$ under the side conditions $\Sigma c_i b_{ii}^2 = 1$ and $D_1(q, Q) = 0$, hence negative since 0 is no eigenvalue. This means that $D_2(q, Q) < 0$ unless all b_{ii} vanish, and proves (7.18).

Alexandrov deduces in [1, IV] from (7.17) that a regular convex hypersurface is determined up to translations when one of the functions $\{R_1 \ldots R_m\}$, $1 \leqq m \leqq n-1$, is known as function of the normal w. We will prove later a corresponding theorem on general convex hypersurfaces.

We now turn to the *proof of* (7.4). Since every convex body can be approximated by regular convex bodies, see K., p. 36, and the mixed volumes depend continuously on the bodies, it suffices to establish (7.4) for regular bodies.

If $N_i(v)$, $i = 1, \ldots, n$, is a positive homogeneous function of degree 1 and of class C^2 for $v \neq 0$ we can formally define

$$V(N_1, \ldots, N_n) = n^{-1} \int_Z N_1(w) D(N_2, \ldots, N_{n-1}, w) d\sigma$$

in analogy to (6.9). It remains true that $V(N_1, \ldots, N_n)$ is independent of the order of the N_i. This is readily verified, but follows from the case of convex functions where we actually apply this notion. It is convenient to denote the supporting function of the convex bodies $H_1, \ldots, H_{n-1}$ by $H_1(v), \ldots, H_{n-1}(v)$. We then show that for regular bodies $H_1, \ldots, H_{n-2}$ and any N which is the difference of two convex functions of class C^2: [1]

$$(7.25) \quad V^2(H_1, \ldots, H_{n-1}, N) \geqq V(H_1, \ldots, H_{n-1}, H_{n-1}) V(H_1, \ldots, H_{n-2}, N, N)$$

with the equality sign only when $N(v) = \lambda H_{n-1}(v) + av$, $\lambda \geqq 0$, (where $av = a \cdot v = \Sigma a_i v_i$), i.e., when $N(v)$ is the supporting function of a convex body homothetic to H_{n-1}.

Putting

$$V_k(H, N) = V(H_1, \ldots, H_{n-k}, \underbrace{N, \ldots, N}_{k})$$

[1] Actually this means any N of class C^2, see A. [7].

we see, as (7.17) was deduced from (7.18), that (7.25) is equivalent to

(7.26) $$V_1(H, N) = 0 \text{ implies } V_2(H, N) \leqq 0,$$

with equality only when $N = av$, i.e., is the supporting function of a point.[2]

We want to apply our results on mixed discriminants to the expressions $D(H^{(1)}, \ldots, H^{(n-1)}, w)$ of Section 6. But the $H^{(j)}(w)$ are functions of n (instead of $n - 1$) variables. This difficulty is overcome by using what we will call *standard coordinates* which vary from point to point: if w on Z is given we choose coordinates $u_1, \ldots, u_n$ with the same origin and orientation such that w becomes the point $u_1 = \ldots = u_{n-1} = 0$ and $u_n = 1$. If $u_n > 0$, then since $\partial H^{(j)}/\partial u_k$ is positive homogeneous of degree 0,

$$\partial H^{(j)}(0, \ldots, 0, u_n)/\partial u_k = \partial H^{(j)}(0, \ldots, 0, 1)/\partial u_k.$$

Therefore $\partial^2 H^{(j)}(0, \ldots, 0, 1)/\partial u_k \partial u_n = 0$ and the second differential of $H^{(j)}$ becomes

$$\sum_{i,k=1}^{n-1} \frac{\partial H^{(j)}}{\partial u_i \, \partial u_k} du_i \, du_k = \sum_{j,k=1}^{n-1} H_{ik}^{(j)} \, du_i \, du_k$$

With these coordinates the expression $D(H^{(1)}, \ldots, H^{(n-1)}, u)$ in Section 6 becomes exactly the mixed discriminant of the second differentials above, so that

$$D(\lambda_1 H^{(i)} + \ldots + \lambda_{n-1} H^{(n-1)}, u) \\ = \sum \lambda_{i_1} \ldots \lambda_{i_{n-1}} D(H^{(i_1)}, \ldots, H^{(i_{n-1})}, u).$$

The geometric meaning of the left side, see (6.11), shows that the discriminants on the right are independent of the coordinates, so that $D(H^{(1)}, \ldots, H^{(n-1)}, w)$ may, for each w, be evaluated by taking the standard coordinates u at w, and this is always understood to be done in what follows.

We return to our previous notation replacing $H^{(i)}$ by H_i and using abbreviations like $D_1(H, N, w) = D_1(H_1, \ldots, H_{n-2}, N, w)$.

[2] That $\lambda \geqq 0$ in (7.25) follows afterwards from its definition and the fact that mixed volumes of convex bodies are non-negative.

We also choose the origin in the interior of $H_1, \ldots, H_{n-1}$ so that $H_i(v) > 0$ for $v \neq 0$. *Hilbert's method* (whose analogue we used already for (7.18)) consists in reducing (7.4) to the following variational problem: *To find the extrema of*

$$V_2(H, N) = n^{-1} \int_Z N(w)\, D_1(H, N, w) d\sigma$$

under the condition

$$n^{-1} \int_Z D_1(H, H_{n-1}, w)\, H_{n-1}^{-1}(w)\, N^2(w)\, d\sigma = 1.$$

Since the bodies H_i are regular, the second differentials of the H_i are positive definite, hence $D_1(H, H_{n-1}, w) > 0$, see (7.13). The Euler equation of this problem is

$$D_1(H, N, w) + \lambda N\, D_1(H, H_{n-1}, w)\, H_{n-1}^{-1}(w) = 0, \tag{7.27}$$

where λ is the eigenvalue parameter.

This is a self-adjoint linear elliptic differential equation on the unit sphere. The linearity is obvious and the equation is self-adjoint because

$$\int_Z M(w)\, D_1(H, N, w) d = \int_Z N(w)\, D_1(H, M, w) d\sigma.$$

The elliptic character is derived from the fact that in suitable standard coordinates (i.e., with a suitable choice of $u_1, \ldots, u_{n-1}$)

$$D_1(H, N, u) = \sum_i D(H_1, \ldots, H_{n-2}, u)_{ii} \frac{\partial^2 N(u)}{\partial u_i^2},$$

where we use the previous notations, and $D(H_1, \ldots, H_{n-2}, u)_{ii} > 0$. It follows from Hilbert's theory, see Hilbert [1], Chapter 18, that the eigenvalues are discrete and bounded from below. We show first

(7.28) *The equation* (7.27) *has* 0 *as eigenvalue with* $N(w) = aw$ *as eigenfunctions, so that* 0 *has multiplicity* n.

That 0 is eigenvalue and aw are eigenfunctions is obvious. To see that there are no others we observe that $D_1(H, N, u) = 0$ implies by (7.18) that $D_2(H, N, u) \leqq 0$. Multiplying the last relation by $H_{n-2}(w) > 0$ and integrating over Z gives

$$V_2(H, N) \leqq 0.$$

On the other hand, multiplying $D_1(H, N, w) = 0$ by $N(w)$ and integrating over Z yields $V_2(H, N) = 0$, hence $D_2(H, N, w) \equiv 0$; and by (7.18) all second derivatives of $N(w)$ vanish in standard coordinates, hence also in other coordinates and $N(w)$ is linear.

(7.29) *The equation* (7.27) *has* $\lambda = -1$ *as only negative eigenvalue (of multiplicity* 1) *with eigenfunction* $N(w) = H_{n-1}(w)$.

Again it is clear that $\lambda = -1$ is an eigenvalue and $H_{n-1}(w)$ an eigenfunction. The argument is similar to that for (7.21):

If $H_1, \ldots, H_{n-1}$ are the unit sphere U, then (7.27) becomes the Laplace equation for $N(u)$ on Z. This follows from $H_{n-1}(w) \equiv 1$ and from the relations (6.11) and (6.12). It is known that this equation has -1 as only negative eigenvalue with $N(w) = 1$ as eigenfunction, see Courant Hilbert [1], p. 511. We consider the bodies $H_i^\rho = (1 - \varrho) U + \varrho H_i$, $i = 1, \ldots, n - 1$, and the corresponding equation

$$D_1(H^\rho, N, w) = \lambda N\, D_1(H^\rho, H_{n-1}{}^\rho, w)[H_{n-1}{}^\rho(w)]^{-1} = 0.$$

As ϱ varies from 0 to 1 the eigenvalues vary continuously, see Hilbert [1] or Courant-Hilbert [1], Chapter VI. By (7.28) 0 always remains an eigenvalue of multiplicity n, hence there is for all ϱ exactly one negative eigenvalue, so that $\lambda = -1$ is the only one for $\varrho = 1$.

The remainder of the proof runs again like that of (7.18). We write the hypothesis $V_1(H, N) = 0$ as

$$n^{-1} \int_Z H_{n-1}{}^{-1}(w)\, D_1(H, H_{n-1})\, H_{n-1}(w) N(w) d\sigma = 0.$$

It then expresses the orthogonality of $N(w)$ to $H_{n-1}(w)$. Because the eigenvalues of (7.27) have the opposite sign from the corresponding extrema, we conclude that $V_2(H, N) \leqq 0$ if $V_1(H, N) = 0$, and that the equality sign holds only when $N(w)$ is an eigenfunction belonging to 0, i.e., when it is linear. This proves (7.25) and (7.4).

That the conditions for the equality in (7.4) for non-regular bodies are different follows from our discussion of the equality sign in (7.3). We also noticed that the equality holds in (7.4) if the function $f(\varrho)$ of (7.5) is linear. Hence $f(\varrho)$ is not linear when $K, L, C_1, \ldots, C_{n-m}$ are regular convex bodies and K and L are not homothetic. *This establishes* (7.10) *in the case where all bodies are regular.*

8. Minkowski's problem

In its simplest form Minkowski's problem asks whether, given a smooth positive function $f(w)$ on the unit sphere Z in E^3 satisfying $\int_Z w^i f(w) d\sigma = 0$, $i = 1, \ldots n$, there is a, up to translations unique, closed convex surface with $f^{-1}(w)$ as Gauss curvature. Minkowski solved this problem (with methods extending to E^n) in a certain generalized sense, (discussed in K., Section 13) and also, appropriately modified, for polyhedra. This led to the question of whether the problem with a suitable substitute for $f(w)$ in terms of set functions can be solved for general closed convex hypersurfaces. Alexandrov [1, I, II] and Fenchel-Jessen [1] showed, independently, that this is indeed the case, so that we have here a *first example of a deeper theorem of differential geometry in the large proved for a geometrically natural class of surfaces*, i.e., without smoothness requirements necessitated by the methods rather than the problem.

The generalization of $f(w)$ used in the two approaches differ slightly: Alexandrov uses methods analogous to those leading to his theorem (4.6), but the sets where his function is defined do not, in general, comprise all Borel sets. Fenchel and Jessen define the set function implicitly on all Borel sets, but give later an explicit interpretation. We follow the latter method which is exceptionally elegant.

Some auxiliary facts on set functions on the unit sphere Z in E^n are needed, for which the reader may consult Radon [1] or A [6] (English). Since we use in this and the next section no set functions other than *completely additive non-negative set functions defined on all Borel sets* ν' *of* Z, we will denote these briefly as *Set functions*. The integral of a function $H(w)$ over ν' with respect to a set function $F(\nu')$ is denoted by

$$\int_{\nu'} H(w)\ F(dZ).$$

(8.1) *If F and G are Set functions and*

$$\int_Z H(w)\ F(dZ) = \int_Z H(w)\ G(dZ)$$

for every supporting function of a convex body, then

$$F \equiv G, \text{ i.e., } F(\nu') = G(\nu') \textit{ on all Borel sets } \nu'.$$

This theorem is well known if $H(w)$ is allowed to traverse all continuous functions on Z. Therefore it suffices to see that a given continuous function $k(w)$ can be uniformly approximated by linear combinations of supporting functions. Because we may write $k(w)$ as the difference of two non-negative continuous functions it suffices to take $k(w) \geqq 0$.

For a fixed $u \in Z$ let $H_{u,\delta}(w)$ be the supporting function of the convex closure of U and δu, $\delta > 1$. Then

$$H_{u,\delta}(w) \begin{cases} > 1 \text{ for all } w \text{ with spherical distance} < \text{arc cos } \delta^{-1} \text{ from } u \\ = 1 \text{ otherwise.} \end{cases}$$

Clearly

$$J_\delta^{-1} = \int_Z (H_{u,\delta}(w) - 1) d\sigma$$

is independent of u. Then

$$F_\delta(u) = \int_Z k(w)[H_{u,\delta}(w) - 1] J_\delta d = \int_Z kH_{u,\delta} J_\delta d\sigma - \int_Z kJ_\delta d\sigma$$

tends for $\delta \to 1$ uniformly to $k(u)$ and the two integrals on the right are, as continuous linear combinations with $k(w) \geqq 0$ of supporting functions, themselves supporting functions, see K., p. 28.

A Borel set ν' is a *continuity set* of $F(\nu')$ if the values of F for ν' and the open kernel of ν' are equal. *If $F(\nu') \leqq G(\nu')$ for all common continuity sets of F and G then this inequality holds for all Borel sets.* Consequently, $F \equiv G$ if $F(\nu') = G(\nu')$ for all common continuity sets.

The sequence $F_1, F_2, \ldots$ of Set functions converges (by definition) if there is a Set function F such that

$$F_i(\nu') \to F(\nu') \textit{ for every continuity set } \nu' \textit{ of } F.$$

By the preceding remarks F is uniquely determined and we write $F_i \to F$. Since $F_i \geqq 0$ it suffices for $F_i \to F$ that $F_i(Z) \to F(Z)$ and lim inf $F_i(\nu') \geqq F(\nu')$ for all open ν'.

(8.2) *The sequence $F_1, F_2, \ldots$ of Set functions converges if*

$$\lim_{i\to\infty} \int_Z H(w)\, F_i(dZ)$$

exists for every supporting function $H(w)$ and only if it exists for every continuous $H(w)$. If $F_i \to F$ then

$$\lim_{i\to\infty} \int_Z H(w)\, F_i(dZ) = \int_Z H(w)\, F(dZ)$$

for every continuous $H(w)$ and this convergence is uniform on any set of equicontinuous $H(w)$.

Everything in this theorem is known except the weakening of the sufficiency condition from continuous $H(w)$ to supporting $H(w)$. This follows by the same approximation method as above.

We are now ready to define the analogue to the area $\int_{\nu'} K^{-1}(w) d\sigma$ of the inverse image of ν' under the spherical mapping of a hypersurface with positive curvature. With the notation of Section 6, let P be a non-degenerate convex polyhedron with $(n-1)$-dimensional faces $p^1, \ldots, p^N$, and denote by w^i, $v(p^i)$ the exterior unit normal and the area of p^i. The area function $A(P, \nu')$ of P is, for a given Borel set ν', defined as the sum of those $v(p^i)$ for which $w^i \in \nu'$. This is evidently a Set function and for any convex body H with supporting function $H(w)$, see (6.4),

$$V_1(P, H) = n^{-1} \sum_{i=1}^{N} H(w^i) v(p^i) = n^{-1} \int_Z H(w)\, A(P, dZ).$$

If K is an arbitrary convex body and $P_1, P_2, \ldots$ a sequence of convex polyhedra tending to K, then

$$\lim V_1(P_i, H) = \lim n^{-1} \int_Z H(w)\, A(P_i, dZ) = V_1(K, H).$$

Because of (8.2) *there exists exactly one Set function $A(K, \nu')$, called the area function of K, such that for every convex body H*

$$V_1(K, H) = n^{-1} \int_Z H(w)\, A(K, dZ). \tag{8.3}$$

Since the mixed volumes are invariant under translations the area function is too. For $H = U$ the mixed volume (8.3) is the area $A(K)$ of ∂K, see Section 6, so that $A(K, Z) = A(K)$.

If $K_i \to K$ then it follows from (8.2) that $A(K_i, \nu')$ converges to a set function $F(\nu')$ satisfying

$$V_1(K, H) = n^{-1} \int_Z H(w)\, F(dZ)$$

for every convex body H, therefore $F(\nu') = A(K, \nu')$. Thus:

(8.4) $\qquad A(K_i, \nu') \to A(K, \nu')$ *if* $K_i \to K$.

We conclude from (6.8) and (6.11) that

$$A(K, \nu') = \int_{\nu'} R_1 \ldots R_n \, d\sigma \text{ for regular } K,$$

so that the area function is a true generalization of $K^{-1}(w)$. If H is a point a then

$$0 = V_1(K, a) = n^{-1} \int_Z aw\, A(K, dZ)$$

therefore

$$\int_Z w_i\, A(K, dZ) = 0,\ i = 1, \ldots, n.$$

The dimension of K *is less than* $n - 1$ *if, and only if,* $A(K) = 0$, *hence* $A(K, \nu') \equiv 0$. If $\dim K = n - 1$, then K lies in a hyperplane with normals w' and $-w'$, say. Then $A(K, \nu') = 0,\ = A(K)/2,\ = A(K)$ according to whether ν' contains none, one, or both of the points w' and $-w'$. In these two cases the area function of K does, therefore, not determine K up to translations.

It will prove useful to reformulate the conditions for these two cases as follows:

(8.5) $\dim K = n$ *if, and only if,* $A(K, \nu_u) < A(K, Z) = A(K)$ *for the intersection* ν_u *of* Z *with an arbitrary hyperplane* $u \cdot x = 0$, $|u| = 1$.

The "only if" is obvious: If $\dim K < n$ then K lies in a hyperplane with normals w', $-w'$, say, and $A(K, w') + A(K, -w') = A(K)$, hence $A(K, \nu_u) = A(K)$ for any plane $u \cdot x = 0$ containing w'.

Assume $A(K, \nu_u) = A(K, Z)$ for some u, then $A(K, \nu') = 0$ for

$\nu' \cap \nu_u = 0$. The segment T_u from $z = (0, \ldots, 0)$ to u has

$$T_u(w) = \max (0, w \cdot u)$$

as supporting function, see K., p. 25, which vanishes for $w \in \nu_u$. Therefore

$$V_1(K, T_u) = n^{-1} \int_Z T_u(w)\, A(K, dZ) = 0.$$

On the other hand $nV_1(K, T_u)$ is the area of the projection of K on $u \cdot x = 0$, see K., p. 45, hence K lies in a hyperplane.

For regular K and for polyhedra we have an explicit expression for $A(K, \nu')$. *For general* K, Fenchel and Jessen give the following *explicit expression*: For a given supporting plane P_u of K with unit normal u form the set $P_u \cap K + h\, T_u$, $h > 0$, obtained by laying off the segment hT_u from every point of $P_u \cap K$. Then for any Borel set ν' on Z

$$A(K, \nu') = \lim_{h \to 0+} h^{-1} V[\bigcup_{u \in \nu'} (P_u \cap K + hT_u)].$$

We do not prove this because we do not need it.

This is now our solution of Minkowski's Problem:

(8.6) THEOREM. *A non-negative completely additive set function $F(\nu')$ defined on all Borel sets ν' of the unit sphere Z of E^n, which satisfies*

(a) $\int_Z w_i\, F(dZ) = 0,\ i = 1, \ldots, n,$

and for every $u \in Z$

(b) $F(\nu_u) < F(Z)$, *where* $\nu_u = (u \cdot x = 0) \cap Z$,

is the area function of one, and up to translations only one, (n-dimensional) convex body in E^n.

The uniqueness follows immediately from (7.2): If H and K have $F(\nu')$ as area function, then (b) and (8.5) yield that $\dim H = \dim K = n$ and (8.3) implies

$$V_1(K,H) = n^{-1} \int_Z H(w)\, A(K,dZ) = n^{-1} \int_Z H(w)\, A(H, dZ) = V(H)$$

and similarly $V_1(H, K) = V(K)$.

This and (7.2) yield

$$V^n(H) = V_1{}^n(K, H) \geqq V^{n-1}(K)V(H),\ V^n(K) \geqq V^{n-1}(H)V(K),$$

hence $V(H) = V(K)$ and the equality sign holds in these inequalities. By (7.2) the bodies H and K are homothetic, and originate from each other by a translation because they have the same volume.

For the existence proof we first take the case where a finite set ν' on Z exists for which $F(\nu') = F(Z)$. If $w^1, \ldots, w^N$ are the points of ν' we may assume $F(w^j) = v^j > 0$. Then (a) and (b) imply that

$$\sum_{j=1}^{N} w^j v^j = 0$$

and that there are n linearly independent w^j. Minkowski proved (for $n = 3$, but his methods extend to $n > 3$, see K., p. 118) that there is a convex polyhedron P with $A(P, \nu') = F(\nu')$. A proof of this fact based on topological methods is found in A. [8], pp. 93–95 and Chapter VII.

From this fact the *general case is obtained by approximation with polyhedra*. We decompose Z into a finite number of Borel sets with diameter less than i^{-1} and denote by $\nu^1, \ldots, \nu^N$ those among these sets for which F is positive. Denote by $\varrho_j w^j$, $|w^j| = 1$, $\varrho_j > 0$, the center of gravity of ν^j with the mass distribution given by $F(\nu')$, i.e., we define w^j and ϱ_j by

$$\int_{\nu^j} w_i{}^j\, F(dZ) = F(\nu^j)\varrho_j w_i{}^j,\ i = 1, \ldots, n;\ j = 1, \ldots, N.$$

Then arc cos $i^{-1} < \varrho_j \leqq 1$ so that ϱ_j, which depends on i, tends uniformly to 1 when $i \to \infty$. Define $F_i(\nu')$ as the sum of those $F(\nu^j)\varrho_j$ for which $w^j \in \nu'$. Then $F_i(\nu')$ satisfies (a); if $H(w)$ is any continuous function on Z and $H_i(w)$ is the function which has on ν^j the value $H(w^j)\varrho_j$, $j = 1, \ldots, N$ and equals $H(w)$ on $Z - \cup \nu^j$ then

$$\int_Z H(w)\ F_i(dZ) = \int_Z H_i(w)\ F(dZ).$$

Also, $H_i(w)$ tends uniformly to $H(w)$ because ϱ_j tends uniformly to 1. Therefore

$$\lim_{i\to\infty}\int_Z H(w)\, F_i(dZ) = \int_Z H(w)\, F(dZ), \tag{8.7}$$

and it follows from (8.2) that $F_i \to F$.

For an arbitrary $u \in Z$ we consider again the segment T_u from 0 to u and its supporting function $T_u(w) = \max\ (0, w \cdot u)$, which is positive on the open hemisphere Z_u of Z with center u and vanishes on the rest of Z. Then

$$\int_Z T_u(w)\, F(dZ) > 0,$$

because, otherwise $F(Z_u) = 0$ and (a) would yield $F(Z_{-u}) = 0$ so that $F(Z) = F(\nu_u)$ in contradiction to (b). The last integral depends continuously on u, hence

$$\int_Z T_u(w)\, F(dZ) \geqq 2c > 0 \text{ for all } u \text{ and a suitable } c.$$

By (8.2), since the $T_u(w)$, $u \in Z$, are trivially equicontinuous,

$$\lim_{i\to\infty}\int_Z T_u(w)\, F_i(dZ) = \int_Z T_u(w)\, F(dZ)$$

uniformly in u, whence with a suitable i_0

$$\int_Z T_u(w) F_i(dZ) > c \text{ for } i \geqq i_0 \text{ and all } u.$$

It follows that F_i satisfies condition (b) for $i \geqq i_0$ because we know that $F_i(\nu_u) = F(Z)$ would imply that the last integral vanishes.

According to our result on the first case there exists a polyhedron P_i, $i \geqq i_0$, with F_i as area function, and owing to the invariance of the area functions under translations we may assume that P_i contains the origin z in its interior. By (8.7) the area $A(P_i) = F_i(Z)$ of P_i tends to $F(Z)$ and is therefore bounded. Applying (7.2) to $L = U$ yields the isoperimetric inequality

$$V^{n-1}(K) \leqq \pi_n A^n(K),$$

so that the volumes of the P_i are also bounded. If the segment βT_u, $\beta > 0$, lies in P_i then the supporting function of P_i is at least $\beta T_u(w)$, therefore

$$n\,V(P_i) \geqq \beta \int_Z T_u(w)\,F_i(dZ) > c\beta$$

and

$$P_i \subset U(z, c^{-1} \sup_i V(P_i)).$$

Consequently a suitable subsequence $\{P_k\}$ of $\{P_i\}$ will converge to a convex body K and by (8.4)

$$F_k(\nu') = A(P_k, \nu') \to A(K, \nu').$$

It follows that $F(\nu') = A(K, \nu')$ which proves (8.6).

We conclude from theorem (5.6) that for $n = 3$ *the surface* ∂K *is of class* C^{m+1} *if* $F(\nu') = \int_{\nu'} f(w)d\sigma$ *and* $f(w)$ *is positive and of class* C^m, $m \geqq 2$.

We also notice that (8.6) contains for $n = 3$ that *a sphere is uniquely determined by its intrinsic metric*, after it has been established that the extrinsic curvature $\nu(M)$ is intrinsic.

9. Uniqueness for given $\{R_1 \ldots R_m\}$

An analogue to the existence statement in theorem (8.6), when curvature functions $\{R_1 \ldots R_m\}$ with $m < n - 1$ are given, is *not known, only the uniqueness has been proved.* The difficulty does not derive from the generality of the hypersurfaces admitted, it is this: If $f(w) = \{R_1 \ldots R_m\}$ then $\int_Z w_i f(w)d\sigma = 0$, $i = 1,\ldots, n$, see (6.15); it is known that $f(w)$ *must, if* $m < n - 1$, *satisfy other conditions in order to come from a convex hypersurface,* but so far there is not even a plausible conjecture what these might be. [3]

That other conditions are necessary was discovered by Alexandrov [1, III]. The simplest case is $n = 3$, $f(w) = R_1 + R_2$. On the unit sphere Z in E^3 we introduce polar coordinates α, β by

$$w_1 = \sin\alpha\cos\beta,\ w_2 = \sin\alpha\sin\beta,$$
$$w_3 = \cos\alpha,\ 0 \leqq \alpha \leqq \pi,\ 0 \leqq \beta < 2\pi.$$

For a, not necessarily convex, closed surface containing z in

[3] A sufficient but not necessary condition in the simplest case is found in Pogorelov [6].

its interior we define $H(w) = h(\alpha, \beta)$ as the distance from z of a tangent plane with exterior normal u. Then (6.11) still holds and, see for example Courant-Hilbert [1], p. 225, and K., p. 66.

$$(9.1) \quad R_1 + R_2 = H_{11} + H_{22} + H_{33} = \frac{1}{\sin\alpha}\frac{\partial}{\partial\alpha}\left(\sin\alpha\frac{\partial h}{\partial\alpha}\right) + \frac{1}{\sin^2\alpha}\frac{\partial^2 h}{\partial\beta^2} + 2h.$$

For a surface of revolution about the w_3-axis $h(\alpha, \beta)$ depends only on α and the second term on the right drops out. In that case the radius of curvature of the meridian is

$$R = h + d^2h/d\alpha^2.$$

In particular, if

$$h(\alpha, \beta) = 3/2 - \cos^2\alpha,$$

$$\text{then } R_1 + R_2 = 4\cos^2\alpha + 1, \ R = 3\cos^2\alpha - \tfrac{1}{2}.$$

Consequently $R_1 + R_2 > 0$, but R changes sign, so that *the surface is not convex*. Although this surface possesses no first differential for the values of α where $R = 0$, it is, up to translations, the only surface with $R_1 + R_2 = 4\cos^2\alpha + 1$ because h must satisfy (9.1), which has an up to a linear function $a \cdot w$ unique solution, see K., p. 124. [4]

Alexandrov (in [1, III]) investigated the situation more closely and constructed *positive analytic functions $f(w)$ on the unit sphere Z in E^n with $\int w_i f(w) d\sigma = 0$ which are not the curvature function $\{R_1 \ldots R_m\}$ of a convex hypersurface for any $m < n - 1$*. The gist of his argument is this: if $F(v')$ is the area function of a convex hypersurface corresponding to $\{R_1 \ldots R_m\}$ with $m < n - 1$ — a rigorous definition will be given presently — then he shows that a point $w \in Z$ with $F(w) > 0$ cannot exist. On the other

[4] It has been proved (Christoffel for $n = 3$, Kubota for arbitrary n) that for a given smooth $f(w)$ with $\int_Z w_i f(w) d\sigma = 0$ a, up to translations unique, closed hypersurface exists for which $f(w) = R_1 + \ldots + R_n$; see K., pp. 123, 124, where it is erroneously stated that this hypersurface is convex.

hand, given a Set function $F(\nu')$ satisfying $\int_Z w_i\, F(dZ) = 0$ he constructs a sequence of positive analytic functions $f_j(w)$ with $\int_Z w_i f_j(w) d\sigma = 0$ such that the Set function

$$F_j(\nu') = \int_{\nu'} f_j(w) d\sigma \to F(\nu').$$

If $f_j(w)$ is the curvature function $\{R_1 \ldots R_m\}$, $m < n - 1$, of a convex body K_j and $K_j \to K$ then we will see that $F_j \to F$, in analogy to (8.4), and that F is the corresponding area function of K. This is impossible if $F(w) > 0$ for a point w, hence $f_j(w)$ can in that case (for large j) not be a function $\{R_1 \ldots R_m\}$ with $m < n - 1$.

From the discussions in Section 6 we expect that the area functions corresponding to $\{R_1 \ldots R_m\}$ with $m < n - 1$ are obtained by considering suitable mixed volumes. We prove first

(9.2) *If $n - 1$ convex bodies $K_1, \ldots, K_{n-1}$ in E^n are given then there exists exactly one Set function $A(K_1, \ldots, K_{n-1}, \nu')$, called the mixed area function of $K_1, \ldots, K_{n-1}$, such that for any convex body H with supporting function $H(w)$*

$$V(H, K_1, \ldots, K_{n-1}) = n^{-1} \int_Z H(w) A(K_1, \ldots, K_{n-1}, dZ).$$

Moreover, $A(K_1, \ldots, K_{n-1}, \nu')$ depends continuously on $K_1, \ldots, K_{n-1}$ and equals $A(K, \nu')$ if $K_i = K$.

The proof is strictly analogous to that of (8.3) and (8.4): We consider first the case where the K_i are polyhedra P_i. Using the notations leading to (6.5) we define $A(P_1, \ldots, P_n, \nu')$ as the Set function whose value for a given Borel set ν' equals the sum of those $v(p_1{}^i, \ldots, p_{n-1}{}^i)$ for which $w^i \in \nu'$. Then, see (6.5),

$$V(H, P_1, \ldots, P_{n-1}) = n^{-1} \sum_{n=1}^{N} H(w') v(p_1{}^i, \ldots, p_{n-1}{}^i)$$

$$= n^{-1} \int_Z H(w) A(P_1, \ldots, P_{n-1}, dZ).$$

The remaining arguments are identical to those for (8.3) and (8.4). We deduce from (6.9) that for sufficiently smooth ∂K_i

$$(9.3) \qquad A(K_1, \ldots, K_{n-1}, \nu') = \int_{\nu'} D(H_1, \ldots, H_{n-1}, w)d\sigma,$$

where H_i is the supporting function of K_i.

In analogy to previous notations we put

$$A_m(K, \nu') = A(\underbrace{K, \ldots, K}_{m}, \underbrace{U, \ldots, U}_{n-m-1}, \nu'), \; 0 \leqq m \leqq n-1,$$

so that

$$A_{n-1}(K,\nu') = A(K,\nu') \text{ and } A_m(K, Z) = nV_{n-m}(K, U) = nV_m(U,K).$$

For smooth K we have because of (9.3), (6.12), and (6.13)

$$(9.3) \qquad A_m(K, \nu') = \binom{n-1}{m} \int_{\nu'} \{R_1 \ldots R_m\} d\sigma.$$

We call $A_m(K, \nu')$ the mth *area function* of K and prove the following uniqueness theorem:

(9.4) THEOREM. *Two convex bodies of dimension at least $m + 1$ have the same* mth *area function if, and only if, one can be carried into the other by a translation.*

The invariance of mixed volumes under translation implies the same for the area functions, so that the sufficiency is obvious. The uniqueness follows from (7.8) and (7.10) in the same way as (8.6) followed from (7.2):

If $\dim H \geqq m + 1$, $\dim K \geqq m + 1$, and $A_m(h, \nu') = A_m(K, \nu')$ then, on account of (9.11), with the notations introduced after (7.4) and $C_i = U$

$$V_{m+1, m}(C, H, K) = n^{-1} \int_Z H(w) A_m(K, dZ) = n^{-1} \int_Z H(w) A_m(H, dZ)$$
$$= V_{m+1, 0}(C, H, K) = V_{m+1}(U, H).$$

Similarly

$$V_{m+1}(U, K) = V_{m+1, m}(C, K, H) = V_{m+1, 0}(C, K, H)$$

and by (7.8)

$$V_{m+1}^{m+1}(U, H) = V_{m+1, m}^{m+1}(C, H, K) \geqq V_{m+1, 0}(C, K) V_{m+1, m+1}^{m}(C, K, H) =$$
$$= V_{m+1}(U, H) V_{m+1}^{m}(U, K).$$

In the same way

$$V_{m+1}^{m+1}(U, K) \geqq V_{m+1}(U, K)\, V_{m+1}^{m}(U, H).$$

Because H and K have at least dimension $m + 1$ these mixed volumes are positive, see K., p. 41 or Section 6. We conclude that $V_{m+1}(U, H) = V_{m+1}(U, K)$ and that the equality sign holds in these inequalities. By (7.10) the bodies H and K are homothetic. The interpretation of $V_{m+1}(U, K) = V_{n-m-1}(K, U)$ as average of the projections of K on the $(m + 1)$-flats through z, see Section 6, shows that K is congruent to H hence originates from H by a translation.

Using the area functions we can solve the problem of finding among all convex bodies with a given $\int_Z \{R_1 \ldots R_m\} d\sigma > 0$ those which minimize $\int_Z \{R_1 \ldots R_p\} d\sigma$ with $p < m$:

(9.5) THEOREM. *Among all convex bodies K with a given positive* $V(K)$ (corresponding to $n = m$) *or a given* $\int_Z A_m(K, dZ)$, $2 \leqq m \leqq n - 1$, *the sphere and only the sphere minimizes* $\int_Z A_p(K, dZ)$, $1 \leqq p < m$.

The interpretation of $V_m(U, K)$ as projection average shows that $V(K) > 0$ or $\int_Z A_m(K, dZ) > 0$ imply $\int_Z A_p(K, dZ) > 0$. The inequality (7.8) with $C_i = U$ yields

$$V_{m,p}^{m}(C, U, K) \geqq V_{m,0}^{m-p}(U)\, V_m^p(U, K) \text{ or}$$
$$V_p^m(U, K) \geqq V^{m-p}(U)\, V^p(U, K)$$

with equality only when K is homothetic to U, i.e., is a sphere. Because of $A_m(K, Z) = n\, V_m(U, K)$ this is our assertion.

The last inequality, (4.5) and (9.3) furnish the following *isoperimetric inequalities for the integrals* $\int_Z \{R_1 \ldots R_m\} d\sigma$:

$$\left\{\int_Z A_p(K, dZ)\right\}^m \geqq (n\pi_n)^{m-p} \left\{\int_Z A_m(K, dZ)\right\}^p, \; 1 \leqq p \leqq m-1,$$

with equality only for spheres when $\int_Z A_m(K, dZ) > 0$.

$$\left\{\int_Z A_p(K, dZ)\right\}^n \geqq n^n \pi_n^{\,n-p}\, V^p(K), \; 1 \leqq p < n,$$

with equality only for spheres when $V(K) > 0$.

The relations (7.7) yield the further inequalities

$$\left\{\int_Z A_p(K, dZ)\right\}^2 \geqq \int_Z A_{p-1}(K, dZ) \int_Z A_{p+1}(K, dZ).$$

The conditions for the equality sign are not known for general K, but if K is regular then it follows from (7.25) that equality holds only when K is a sphere.

We conclude this chapter by mentioning a result, which is related, in spirit at least, to the results of this section and is due to Hopf and Voss [1] and Voss [1].

Let C and C' be two equally oriented closed convex hypersurfaces with positive curvature of class C^3 in E^n. Assume there is an orientation preserving mapping $p \to p'$ of C on C' such that 1) *the vectors from p to p' are all parallel,* and 2) *One of the elementary symmetric functions $\{R_1^{-1} \ldots R_m^{-1}\} = \{R_1 \ldots R_{n-m-1}\}/R_1 \ldots R_{n-1}$, m fixed, has for C and C' the same value at corresponding points p and p'. Then the mapping $p \to p'$ is (part of) a translation.*

A consequence of this theorem is:

If a straight line L exists such that $\{R_1^{-1} \ldots R_m^{-1}\}$ has the same value at two points of the closed convex hypersurface C of class C^3 with positive curvature which lie on a line parallel to L, then C is symmetric with respect to a hyperplane normal to L.

Many very general uniqueness theorems for smooth surfaces were recently obtained by Alexandrov [11].

CHAPTER III

Intrinsic Geometry

10. Intrinsic metrics

For the intrinsic geometry of convex surfaces some simple facts on general metric spaces are needed which we compile in this section. They are found in many places, in particular in Section II, 1 of A. However, the arguments there can be shortened by proceeding in a different order, see for example Busemann [3], Section 5.

Let R be a metric space with points $x, y, \ldots$ and distance xy. Because the term "metric space" has been generalized in various ways we emphasize that we mean the ordinary conditions $xx = 0$, $xy = yx > 0$ for $x \neq y$, $xy + yz \geqq xz$. A *curve* $x(t)$, $\alpha \leqq t \leqq \beta$, in R is a continuous map of the interval $[\alpha, \beta]$ of the real axis (with distance $| t_1 - t_2 |$) in R. Clearly $x(t)$ is uniformly continuous.

To define the *length* $\lambda(x)$ of $x(t)$ we introduce the symbol Δ_t for a partition $\Delta_t : \alpha = t_0 < t_1 < \ldots < t_n = \beta$ of $[\alpha, \beta]$ and put

$$\| \Delta_t \| = \max_i \, (t_{i+1} - t_i), \quad \lambda(x, \Delta_t) = \sum_{i=1}^{n} x(t_{i-1}) x(t_i).$$

Then, by definition

$$\lambda(x) = \lambda_\alpha^\beta(x) = \sup_{\Delta_t} \lambda(x, \Delta_t).$$

Obviously

$$\lambda_\alpha^\beta(x) \geqq x(\alpha) x(\beta)$$

and we have the usual theorem:

(10.1) *For a given* $\epsilon > 0$ *there is a* $\delta > 0$ *such that* $\| \Delta_t \| < \delta$ *entails*

$$\lambda(x) - \lambda(x, \Delta_t) < \epsilon \text{ if } \lambda(x) < \infty, \; \lambda(x, \Delta_t) > \epsilon^{-1} \text{ if } \lambda(x) = \infty.$$

Curves of finite length are called *rectifiable*. Since we use only such curves, we will, in general, omit "rectifiable."

The following are corollaries of (10.1):

(10.2) *If Δ_t^k is a sequence of partitions of* $[\alpha, \beta]$ *with*

$$\lim_{k\to\infty} \| \Delta_t^k \| = 0, \text{ then } \lambda(x, \Delta_t^k) \to \lambda(x).$$

(10.3) *Additivity of length*: If $\alpha = \alpha_0 < \alpha_1 < \ldots < \alpha_n = \beta$ *and* $\lambda_{\alpha_i}^{\alpha_{i+1}}(x)$ *is the length of* $x(t)$, $\alpha_i \leqq t \leqq \alpha_{i+1}$, *then*

$$\lambda_\alpha^\beta(x) = \sum_{i=1}^{n} \lambda_{\alpha_{i+1}}^{\alpha_i}(x).$$

Very simple applications of these facts yield:

(10.4) *If $x(t)$ is rectifiable then $\lambda_\alpha^t(x)$ is a non-decreasing continuous function of t.*

(10.5) *Lower semicontinuity of length: If $x_k(t)$ and $x(t)$, $\alpha \leqq t \leqq \beta$ are curves in R and $x_k(t) \to x(t)$ for every $t \in [\alpha, \beta]$, then*

$$\liminf \lambda(x_k) \geqq \lambda(x).$$

The parameter s on the curve $y(s)$, $\gamma \leqq s \leqq \delta$ *is arc length* if

$$\lambda_{\gamma'}^{\delta'} = \delta' - \gamma' \text{ for } \gamma \leqq \gamma' < \delta' \leqq \delta.$$

On a given rectifiable curve $x(t)$, $\alpha \leqq t \leqq \beta$, we *introduce arc length as parameter* by defining $y(s)$ for $0 \leqq s \leqq \lambda(x)$ as that point $x(t)$ for which $\lambda_\alpha^t(x) = y(s)$. Notice that the point $x(t)$ is unique, but that t need not be, because $x(t)$ may be constant, i.e., the same point, on a whole t-interval. It is readily deduced from (10.3,4) that s is actually arc length on the curve $y(s)$ thus defined.

The parametrization $x(t)$ of a curve introduces geometrically irrelevant properties. These can be eliminated by considering two rectifiable curves $x_1(t)$ and $x_2(t)$ as *equivalent* when their representations $y_i(s)$, $0 \leqq s \leqq \lambda(x_i)$ in terms of arc length are identical: $y_1(s) = y_2(s)$ for $0 \leqq s \leqq \lambda(x_1) = \lambda(x_2)$.

To illustrate our point we will, for a moment, denote such a class C of equivalent curves as a geometric curve. Every element of C is a parametrization of C, the common length of these parametrizations is the length $\lambda(C)$ of C, etc. That the geometric curves $C_1, C_2, \ldots$ tend (uniformly) to the curve C means: there are parametrizations $x_i(t)$ of C_i and $x(t)$ of C, where always $\alpha \leqq t \leqq \beta$, such that $x_i(t) \to x(t)$ (uniformly) in $[\alpha, \beta]$. The significance of geometric curves becomes clear from the next theorem which is quite important, but not valid for ordinary curves, i.e., without allowing reparametrization.

A metric space R is *finitely compact* if the Bolzano-Weierstrass theorem holds; any bounded infinite set M in R has an accumulation point in R. Boundedness means, of course, that $\sup_{x, y \in M} xy < \infty$ or, equivalently, that $px < \beta(p)$ for any $p \in R$ and $x \in M$.

(10.5′) *If the lengths $\lambda(C_i)$ of the curves C_i, $i = 1, 2, 3, \ldots$, in a finitely compact space are bounded and the initial points of the C_i form a bounded set then $\{C_i\}$ contains a subsequence $\{C_j\}$ which converges uniformly to a curve C.*

It follows from (10.5) *that* $\lambda(C) \leqq \liminf \lambda(C_j)$.

Notice the corollary

(10.6) *If the points u, v of a finitely compact space can be connected by a rectifiable curve, then there exists a shortest join of u and v.*

A curve C is an arc (or Jordan arc) if it possesses a parametrization $x(t)$ with $x(t_1) \neq x(t_2)$ for $t_1 \neq t_2$. [1]

(10.7) *A shortest join of two points is an arc.*

Since $\lambda_\alpha^\beta(x) \geqq x(\alpha)x(\beta)$, a curve $x(t)$ from $a = x(\alpha)$ to $b = x(\beta)$ of length ab (if it exists) is a shortest connection of a and b, hence an arc. We call such an arc a *segment* $T^+(a, b)$ *if oriented going from a to b*. The other orientation is a segment $T^+(b, a)$, and we use $T(a, b)$ for the *non-oriented* arc. The segment $T(a, b)$ need, of course, not be unique, all semi great circles joining two antipodal points on a sphere are segments. $T(a, b)$ need not even be unique in the small: In an (x_1, x_2)-plane with the metric

$$xy = | x_1 - y_1 | + | x_2 - y_2 |$$

any curve $x(t)$ from a to b for which $x_1(t)$ and $x_2(t)$ are monotone

[1] For an arc we will always assume that its parametrization satifies this condition.

is a segment. In intrinsic differential geometry a segment $T(a, b)$ is a shortest geodesic join of a and b.

If $y(s)$, $\alpha \leqq s \leqq \alpha + ab = \beta$ represents a segment $T^+(a, b)$ in terms of arc length then for $\alpha \leqq \gamma < \delta \leqq \beta$

$$ab \leqq ay(\gamma) + y(\gamma)y(\delta) + y(\delta)y(\beta) \leqq \lambda_\alpha^\gamma(y) + \lambda_\gamma^\delta(y) + \lambda_\delta^\beta(y) = \lambda_\alpha^\beta(y) = ab$$

hence

$$\delta - \gamma = \lambda_\gamma^\delta(y) = y(\gamma)y(\delta),$$

so that *every subarc of a segment is a segment and* $y(s) \to s$ *maps* $T^+(a, b)$ *isometrically on the segment* $[\alpha, \beta]$ *of the real axis,* whence the name segment. Clearly by (10.5) if $a_i \to a$, $b_i \to b$ then the limit of any converging subsequence $\{T^+(a_k, b_k)\}$ of $\{T^+(a_i, b_i)\}$ is a segment $T^+(a, b)$.

Extending from E^n to a general metric space R the notation $U(p, \varrho)$ to the set of points x in R with $px < \varrho$ we have the useful lemma

(10.8) *If* $a, b \in U(p, \varrho)$ *then* $T(a, b) \subset U(p, 2\varrho)$.

For, if $x \in T(a, b)$ then $ax + xb = ab$, hence

$$\min_{a, b} (ax, xb) \leqq ab/2 \leqq (ap + pb)/2 < \varrho$$

and if $ax = \min (ax, xb)$ then $xp \leqq xa + ap < 2\varrho$. This estimate seems crude, actually the *factor* 2 *in* (10.8) *is the best possible, even for convex surfaces.* On the unit sphere of E^3 (with its spherical metric) if $\varrho > \pi/2$ then the two points on a great circle through p with distance $\varrho/2 + \pi/4 > \pi/2$ from p lie in $U(p, \varrho)$, but the segment T joining them is unique and passes through the antipodal point to p, so that T will not lie in $U(p, \varrho')$ unless $\varrho' > \pi$.

We introduce the notation (xyz) to indicate that x, y, z are distinct and that $xy + yz = xz$.

(10.9) *If* (wxy) *and* (wyz) *then* (xyz) *and* (wxz) *and conversely.*

For

$$wz = wy + yz = wx + xy + yz \geqq wx + xz \geqq wz.$$

The metric on a circle shows that (wxy) and (xyz) do *not* imply

(wxz) or (wyz). As a corollary of (10.9) we have

(10.10) *If $T(x, y)$ and $T(y, z)$ exist and (xyz) then $T(x, y) \cup T(y, z)$ is a segment $T(x, z)$.*

We follow Alexandrov in calling the metric of R *intrinsic* if any two points of R can be joined by a rectifiable curve and the distance ab equals the greatest lower bound of the lengths of all curves in R joining a to b. The existence of a segment $T(a, b)$ for given points a, b is a sufficient, but not a necessary, condition for intrinsicness.

(10.11) *If R is locally compact and its metric is intrinsic, then every point p of R has a neighborhood $U(p, \varrho)$, $\varrho > 0$, such that any two points in $U(p, \varrho)$ can be joined by a segment.*

For there is a positive ϱ such that the closure $\overline{U}$ of $U(p, 2\varrho)$ is compact. If $a, b \in U(p, \varrho)$ then $ab < 2\varrho$. By hypothesis there is a sequence of curves C_i in R joining a to b and with $\lambda(C_i) \to ab$. Any curve from a to b containing a point c not in $U(p, 2\varrho)$ has at least length $ac + cb \geqq pc - pa + pc - pb > 2\varrho$. Therefore $C_i \subset U(p, 2\varrho)$ for large i. By (10.5′) there is a subsequence $\{C_j\}$ of $\{C_i\}$ which tends to a curve C in $\overline{U}$ and

$$ab \leqq \lambda(C) \leqq \liminf \lambda(C_i) = ab.$$

The curves which are, locally, segments will play a considerable role later and are called geodesics. Explicitly: *a geodesic is rectifiable and if $y(s)$, $\alpha \leqq s \leqq \beta$, is a parametrization of the curve in terms of arc length, then a positive function $\epsilon(s)$ exists such that*

$$\gamma(s_1)\gamma(s_2) = |s_1 - s_2| \text{ if either } |s_i - s| \leqq \epsilon(s) \text{ and } s \neq \alpha, \beta,$$
$$\text{or } \alpha \leqq s_i \leqq \alpha + \epsilon(\alpha) \text{ or } \beta - \epsilon(\beta) \leqq s_i \leqq \beta, \; i = 1, 2.$$

If any two points of a metric space R with distance $\delta(a, b)$ can be connected by a rectifiable curve, then an *intrinsic distance ab* may be defined as the greatest lower bound of the lengths of all curves joining a to b. Clearly, $ab \geqq \delta(a, b)$ and ab satisfies the axioms for a metric space. *However, ab may not be (topologically) equivalent* to $\delta(a, b)$. For example, let C be a closed Jordan curve in a plane P of E_3, which is nowhere differentiable, so that no subarc of C is rectifiable. If q is any point not in P then $S = \bigcup_{x \in C} R(q, x)$ is homeomorphic to a plane, but if a and b lie on S and not on the same ray $R(q, x)$, then $ab = |a - q| + |q - b|$. This example shows that a curve with respect to $\delta(a, b)$ need not be a curve with respect to ab, but a curve with respect to ab is, because of $ab \geqq \delta(a, b)$, a curve for $\delta(a, b)$. That the distance ab is really intrinsic follows from:

(10.12) *If $x(t)$, $\alpha \leqq t \leqq \beta$, is a curve for the intrinsic distance ab of R then its length $\lambda_i(x)$ with respect to ab, equals its length $\lambda(x)$ with respect to the original distance $\delta(a, b)$.*

Obviously $\lambda_i(x) \geqq \lambda(x)$ so that $\lambda(x) = \infty$ entails $\lambda_i(x) = \infty$. If $\lambda(x) < \infty$ and Δ_t is any partition of $[\alpha, \beta]$ then

$$\lambda(x) = \sum_{i=1}^{n} \lambda_{t_{i-1}}^{t_i}(x) \geqq \sum_{i=1}^{n} x(t_{i-1})x(t_i),$$

so that $\lambda(x) \geqq \lambda_i(x)$.

Finally:

(10.13) *Let the metric ab of R be intrinsic. If R' is a connected open subset of R, then its intrinsic metric (ab, R') is locally equal to ab. A geodesic in R' is a geodesic in R.*

For, any two points of R' can be connected by a rectifiable curve since this is the case for R. Let $U(p, 2\varrho) \subset R'$. As in the proof of (10.11), if $\lambda(C_i) \to ab$, then $C_i \subset U(p, 2\varrho)$ for large i, so that $C_i \subset R'$ and $(ab, R') = ab$. The local character of the definition of a geodesic yields the last statement. It should be noticed that a segment in R' need *not* be a segment in R, although it is a geodesic.

11. The metrics of convex hypersurfaces

We apply this general theory first to a complete convex hypersurface C in E^n. As a subset of E^n, it carries the metric $|p - q|$. Any two points of C can be connected by a rectifiable curve on C, for example, by a plane convex curve. Hence *C possesses an intrinsic metric ab, and this is equivalent to $|p - q|$*. For, if we use the notations of (1.12), then $|x - y| \leqq |p - q|$ for two points

$p = (x, f(x))$, $q = (y, f(y))$ of C_b. The boundedness of the difference quotients entails that the arc A from p to q of the plane section of C_b normal to $z = 0$ has length $\lambda(A) \leqq M_b | x - y |$, where M_b is independent of p and q (on C_b). Therefore

$$| p - q | \leqq pq \leqq \lambda(A) \leqq M_b | p - q |,$$

which proves the equivalence of $| p - q |$ and pq.

A complete convex hypersurface C is finitely compact with respect to both $| a - b |$ and the intrinsic distance ab. The former because C is a closed subset of E^n, the latter because $ab \geqq | a-b |$.

By (10.6) *there is a shortest join of two given points a, b on C*, whose length equals ab by definition, and hence is a *segment* $T(a, b)$ with respect to the inner metric ab.

Passing to a proper connected (relatively) open subset C' of C, that is, to a non-complete convex hypersurface, we conclude from (10.14) that the intrinsic metric (ab, C') of C' is locally equal to ab and that a geodesic on C' is a geodesic on C.

Alexandrov proves his principal theorems on the intrinsic metrics of general convex surfaces by approximation with polyhedra. Basic for this procedure is the continuous dependence of the intrinsic metric on the surface. We need the lemma:

(11.1) *Let T_1, T_2 be two distinct half-hyperplanes with a common*

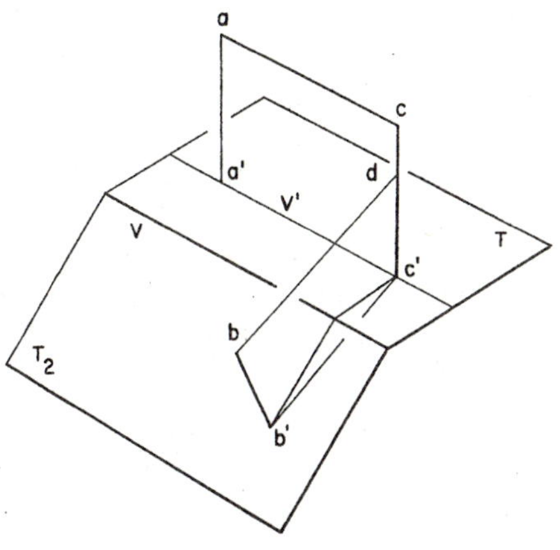

Figure 5

bounding $(n-2)$-flat V and such that $T = T_1 \cup T_2$ is not a hyperplane. If a, b do not lie in the open convex set K bounded by T, then their feet a', b' on K or T satisfy

$$|a' - b'| \leqq |a - b|$$

For a proof we may assume $a' \in T_1 - V$, $b' \in T_2 - V$, because in the remaining cases trivially $|a' - b'| = |a - b|$. Denote by V' the $(n-2)$-flat parallel to V through a'. The two-flat P through b normal to V and V' contains b', intersects V' in a point c' and contains the normal to T_1 at c'. Let c be the point outside K on this normal with $|c - c'| = |a - a'|$. Then

$$|a - b|^2 = |a - c|^2 + |c - b|^2 = |a' - c'|^2 + |c - b|^2.$$

Either the parallel to $L(b', c')$ through b intersects $E(c, c')$ in a point d, or the parallel to $L(b', c')$ through a intersects $E(b, b')$. Take the first case. Because b and a lie outside K, the foot of b on $L(c, c')$ lies on $R(d, c')$ and $|d - b| > |c' - b'|$. Therefore

$$|c - b| \geqq |d - b| > |c' - b'|$$

and

$$|a - b|^2 > |a' - c'|^2 + |c' - b'|^2 = |a' - b'|^2 :$$

This lemma yields a simple, but most useful fact, see B. F. [1]:

(11.2) THEOREM. *Let C be a complete convex hypersurface bounding the open convex set K. If $x(t)$, $\alpha \leqq t \leqq \beta$, $x(\alpha) = a$, $x(\beta) = b$, is a curve outside of K and $x'(t)$ is the foot of $x(t)$ on $\bar{K}$ or C then*

$$\lambda(x) \geqq \lambda(x') \geqq ab.$$

If $a \in C$, $b \in C$ and $x(t)$ contains points not in $C \cup K$, then $\lambda(x) > ab$.

The foot $x'(t)$ of $x(t)$ is unique by (1.7) and therefore depends continuously on t, so that $x'(t)$ is a curve on C. For any point $x(t)$ not on C the hyperplane P_t normal to $L(x(t), x'(t))$ at $x'(t)$ is a supporting plane of C. If Δ_t is a partition of $[\alpha, \beta]$ then $P_{t_{i-1}}$, P_{t_i} either coincide or the intersection of the supporting half space of K bounded by them is a convex set which does not contain $x(t_{i-1})$ or $x(t_i)$ in its interior. The preceding lemma yields:

$$|x(t_{i-1}) - x(t_i)| \geqq |x'(t_{i-1}) - x'(t_i)|, \text{ hence } \lambda(x) \geqq \lambda(x').$$

Let $a \in C$, $b \in C$ and $x(t_0) \notin C$. Because of (1.7) there is a hyperplane P separating $x(t_0)$ from C. Therefore there are values t_1, t_2 with $\alpha < t_1 < t_0 < t_2 < \beta$ such that $x(t_i) \in P$, $i = 1,2$. If C' is the curve originating from $x(t)$ through replacing the subcurve $t_1 \leqq t \leqq t_2$ by $E(x(t_1), x(t_2))$ then by the first part of this proof

$$\lambda(x) > \lambda(C') \geqq ab.$$

When more than one surface is involved we will *denote the intrinsic distances* of two points a, b as points of the convex hypersurface C *by* (ab, C). Convergence of sets we understand, as always, in the sense of Hausdorff's closed limit and prove:

(11.3) THEOREM. *If the complete convex hypersurfaces $C_1, C_2, \ldots$ tend to the hypersurface C and the points a_i, b_i on C_i tend to the points a, b on C respectively, then*

$$(a_i b_i,\ C_i) \rightarrow (ab,\ C).$$

This theorem is found in A., Chapter III, Section 1. We give a somewhat shorter proof. First we consider the case where C and hence C_i for large i, are closed. If C bounds K then $C_i \subset K + U$ for large i. We denote by $A_i(B_i)$ a ray beginning at $a_i(b_i)$ and containing an exterior normal to C_i at $a_i(b_i)$. This ray $A_i(B_i)$ intersects the boundary $\bar{C}$ of $K + U$ in a point $\bar{a}_i(\bar{b}_i)$. Since a segment from $\bar{a}_i$ to $\bar{b}_i$ on $\bar{C}$ lies outside C_i, it follows from (11.2) that

$$\max_{x, y \in \bar{C}} (xy,\ \bar{C}) \geqq (\bar{a}_i \bar{b}_i,\ \bar{C}) \geqq (a_i b_i,\ C_i).$$

If D_i is a segment from a_i to b_i on C_i, choose a subsequence $\{D_j\}$ of $\{D_i\}$ such that

$$\lambda(D_j) \rightarrow \liminf \lambda(D_i).$$

$\{D_j\}$ contains by (10.5) a subsequence $\{D_k\}$ which converges to a curve D. Then $D \subset C$ and

$$(ab, C) \leqq \lambda(D) \leqq \lim \lambda(D_k) = \liminf (a_i b_i, C_i).$$

Next choose a subsequence $\{C_m\}$ of $\{C_i\}$ such that

$$\lim (a_m b_m, C_m) = \lim\sup (a_i b_i, C_i)$$

and then a subsequence $\{C_n\}$ of $\{C_m\}$ such that A_n and B_n tend to rays normal to C at a and b. Place the origin z inside C; for a suitable sequence of numbers α_n with $0 < \alpha_n < 1$ and $\alpha_n \to 1$ the hypersurface $C'_n = \alpha_n C_n$ lies inside C. Then $a'_n = \alpha_n a_n$ and $b'_n = \alpha_n b_n$ tend to a and b, and $\alpha_n A_n$, $\alpha_n B_n$ intersect C in points p_n, q_n with $p_n \to a$, $q_n \to b$ because $\alpha_n A_n$ and $\alpha_n B_n$ tend to rays normal to C at a and b. Therefore $(p_n a, C) \to 0$, $(q_n b, C) \to 0$ and $\lim (p_n q_n, C) = (ab, C)$. We conclude from (11.2) that $(p_n q_n, C) \geqq (a_n b_n, C_n)$, hence

$$(ab, C) \geqq \lim (a_n b_n, C_n) = \lim\sup (a_i b_i, C_i),$$

which proves our assertion for closed C.

The case where C is not closed can be reduced to this case through replacing the convex sets K, K_i bounded by C and C_i by their intersections with a sufficiently large ball $|x| \leqq R$, and C, C_i by the boundaries of these intersections.

For a later application we mention that the theorem remains correct if, in E^3, the limit C of C_i degenerates into a doubly covered plane convex domain, see A., p. 112.

In classical differential geometry the intrinsic metric is locally approximated by that of the tangent plane. For general convex surfaces we expect the local metric to be that of the tangent cone. We make this precise:

(11.4) *Let the, not necessarily complete, convex hypersurface be locally represented in the form $z = f(x)$ as in (1.12). The tangent cone T at the origin b has the equation*

$$z = \lim_{h \to 0+} \frac{f(hx)}{h} = g(x).$$

Put $c_x = (x, f(x)) \in C$, $t_x = (x, g(x)) \in T$. For a given $\epsilon > 0$ there is a $\delta(\epsilon) > 0$ such that

$$|(c_x c_y, C) - (t_x t_y, T)| \leqq \epsilon \max (|x|, |y|), \text{ if } |x| < \delta(\epsilon),\ |y| < \delta(\epsilon).$$

Because of (10.13) we may assume that C is a complete convex hypersurface. That $z = g(x)$ represents T is clear, see also K., p. 20.

The equation

$$z = h^{-1}f(hx), \quad h > 0,$$

represents the surface $h^{-1}C$ which tends for $h \to 0+$ to T. Because in the preceding theorem the way in which a_i and b_i tend to a and b is arbitrary, this theorem contains a uniformity property (explicitly formulated in A., p. 113) which yields the existence of a positive function $\delta(\epsilon)$ such that

(11.5) $\quad 0 < h < \delta(\epsilon)$ and $|h^{-1}x| \leqq 1, \; |h^{-1}y| \leqq 1$

entail

$$\left|(h^{-1}c_x h^{-1}c_y, h^{-1}C) - (t_{h^{-1}x} t_{h^{-1}y}, T)\right| < \epsilon,$$

where we use that $h^{-1}c_x$ and $t_{h^{-1}x} = h^{-1}t_x$ are the points of $h^{-1}C$ and $T = h^{-1}T$ over $h^{-1}x$. But

$$(h^{-1}c_x t_{h^{-1}x} h^{-1}c_y, h^{-1}C) = h^{-1}(c_x c_y, C), \; (t_{h^{-1}x} t_{h^{-1}y}, T) = h^{-1}(t_x t_y, T).$$

Thus under the conditions (11.5)

$$\left|(c_x c_y, C) - (t_x t_y, T)\right| \leqq \epsilon h.$$

If $\max(|x|, |y|) < \delta(\epsilon)$ then (11.5) is satisfied with $h = \max(|x|, |y|)$ and we obtain the assertion.

We notice that (11.4) implies

(11.6) $$\left|(c_x c_y, C) - (t_x t_y, T)\right| \leqq \begin{cases} \epsilon \max\,[(c_x b, C), (c_y b, C)] \; \textit{for} \\ \qquad \max\,[(c_x b, C), (c_y b, C)] < \delta\epsilon, \\ \epsilon \max\,[|t_x - p|, |t_y - p|] \; \textit{for} \\ \qquad \max\,[|t_x - p|, |t_y - p|] < \delta(\epsilon) \end{cases}$$

In this form, which has the advantage that only distances on C and T enter, the theorem is found in A., Chapter IV, Section 5. The mapping $c_x \to t_x$ of C on T depends on the choice of the representation of C. It is important to notice that the theorem may be formulated intrinsically as follows:

(11.7) *There is a topological mapping $x \to x'$ of a neighborhood U of a given point b on a convex hypersurface C on a neighborhood U' of the apex b' of a suitable cone T (with b' as image of b) and a function $\delta(\epsilon) > 0$ such that*

$$|(xy, C)-(x'y', T)| \leqq \begin{cases} \epsilon\,[\max\ (xb, C),\ (yb, C)]\ \textit{for} \\ \qquad\qquad \max\ [(xb, C),\ (yb, C)] < \delta(\epsilon) \\ \epsilon\,[\max\ (x'b', T),\ (y'b', T)]\ \textit{for} \\ \qquad\qquad \max\ [(x'b', T),\ (y'b', T)] < \delta(\epsilon). \end{cases}$$

Actually it is not necessary to write both these inequalities (or both in (11.6)) down, one of them with ϵ implies the other with $\epsilon(1-\epsilon)^{-1}$.

A useful further corollary of (11.4) is:

(11.8) *For any point b on a convex hypersurface C and any sequence p_i on C tending to b*

$$\lim_{i\to\infty} (bp_i, C)/|b-p_i| = 1.$$

For, if with the notation of (11.4), $p_i = (x_i, f(x_i)) = c_{x_i}$ then

$$|b - t_{x_i}|/|b-p_i| \to 1 \text{ and } (bp_i, C)/|b-t_{x_i}| \to 1.$$

We saw in the last section that disks $px < \varrho$ with $\varrho > \pi/2$ on the unit sphere Z in E^3 are not geodesically convex in the sense that they will not always contain a segment $T(a, b)$ when they contain a and b. It is well known that sufficiently small spheres $px < \varrho$ on smooth manifolds are geodesically convex. This is not so on arbitrary complete convex surfaces and the question arises whether every point has small convex neighborhoods. On surfaces we will construct small convex geodesic triangles. On hypersurfaces this method does not work. But caps have many convexity properties:

(11.9) THEOREM. *A convex cap K contains with any two points a, b a segment with respect to the intrinsic metric of the cap.*

If the convex hypersurface C possesses at b a supporting plane P' which touches C only at b then a cap K cut off C by a hyperplane P sufficiently close to P' contains with any two points a, b every segment $T(a, b)$ on C.

The first part of the theorem is found in A., p. 370 for surfaces. Denote by P the hyperplane containing the boundary B of the given cap K. The union of all rays which begin at points of B, are normal to P, and lie on the opposite side of P from K, form together with K a complete open convex hypersurface L.

We prove the first part by showing that a segment T' from a to b on L lies in $\bar{K}$; it suffices to consider $\bar{K}$, also for the second part, because a and b lie in some closed subcap of K. If T' contained a subarc A with end points on B and otherwise in $L - \bar{K}$, then projecting A on P would furnish a curve A' in B which is shorter than A, hence T' was not a shortest connection of a and b on L.

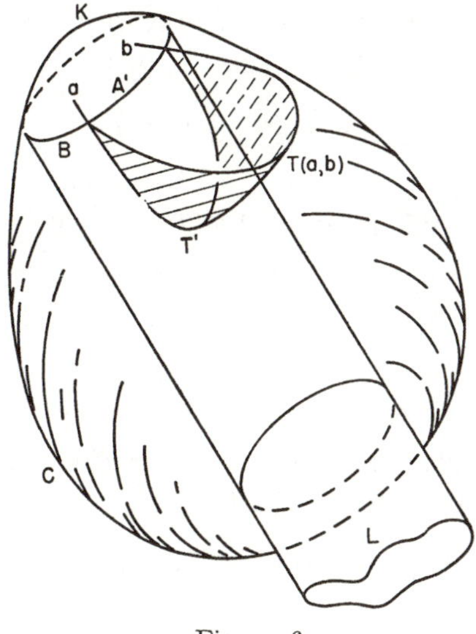

Figure 6

For the second part of (11.9), if P is close to P', then the segments of the generators of $L - K$ inside C are long compared to the distances on K. Therefore the feet on L of a segment $T(a, b)$ on C, $a, b \in K$ will eventually lie inside C. These feet form by (11.2) a curve T' with $\lambda(T') \leqq \lambda(T)$. Points of T not in $\bar{K}$ would have their feet on $L - K$; as in the first part T' could be replaced by a shorter curve from a to b on $\bar{K}$, so that $T(a, b)$ would not be a shortest join of a and b on C.

That large caps of a surface C need not have the property of containing a $T(a, b)$ on C, is seen from the example of a hemisphere H completed by the circular disk with the same boundary as H.

There are other important intrinsic concepts besides distance, in particular, angle and area. The former will be discussed later. An intrinsic area is obtained naturally and easily by using *Hausdorff measure,* whose definition we repeat for convenience.

Consider a metric space R with a countable base. Denote by $\delta(X)$ the diameter of a set X in R (which means $\delta(X) = \sup_{a, b \in X} ab$). For a given $\epsilon > 0$ and a given set M in R we consider all countable coverings $\Lambda : \cup X_i$ of M by sets X_i with $\delta(X_i) < \epsilon$ and put (see (4.5))

$$| M |_m^\epsilon = 2^{-m} \pi_m \inf_{\Lambda} \sum_{i} \delta^m(X_i),$$

where m is any fixed positive integer. Then

$$| M |_m^\epsilon \leqq | M' |_m^{\epsilon'} \text{ if } M \subset M' \text{ and } \epsilon \geqq \epsilon'.$$

Therefore

$$| M |_m = \lim_{\epsilon \to 0+} | M |_m^\epsilon$$

exists and is called the (exterior) *m-dimensional Hausdorff measure* of M. It is trivial that

$$| M |_m \leqq | M' |_m \text{ for } M \subset M', \ | \cup M_i |_m \leqq \sum | M_i |_m.$$

It is also easily seen that

$$| M |_{m'} > 0 \quad \text{entails } | M |_m = \infty \text{ for } m < m' \quad \text{and that}$$

$$| M |_{m'} < \infty \text{ entails } | M |_m = 0 \quad \text{for } m > m'.$$

Thus, in general, $| M |_m$ is significant for only one m.

If C is a complete convex hypersurface with the intrinsic metric xy, then the $(n - 1)$-dimensional Hausdorff measure is significant and is called the *intrinsic area or measure* on C. If C' is a non-complete convex hypersurface and part of the complete hypersurface C, then it follows from (10.13) that the $(n - 1)$-dimensional Hausdorff measure of a set M in C' with respect to the intrinsic metric of C' coincides with the intrinsic area of M as set on C.

The intrinsic measure of a Borel set M on C coincides with the extrinsic measure which we have hitherto used and may therefore be evaluated as follows: we represent M as a countable or finite union of disjoint Borel sets M_i each of which lies in a portion of C representable in the form $z = f_i(x)$ as in (1.12). If M'_i is the projection of M_i on $z = 0$ then

$$|M|_{n-1} = \sum |M_i|_{n-1} \text{ and } |M_i|_{n-1} = \int_{M'_i} \left[1 + \sum_{k=1}^{n-1} \left(\frac{\partial f_i}{\partial x_k}\right)^2\right]^{1/2} dx_1 \ldots dx_{n-1}.$$

Because $(n - 1)$-dimensional Hausdorff measure in E^{n-1} equals Lebesgue measure (this well-known fact is found, for instance, in Busemann [1], p. 239), these facts are easily deduced from (1.2) and (3.1) or also from (11.4). They are very special cases of results in Busemann [1,2]. [2]

Alexandrov defines intrinsic measure only for convex surfaces in E^3 (see A., Chapter X) in a way which is not immediately generalizable to higher dimensions, and which we will briefly outline later (Section 14). The present approach through Hausdorff measure is simpler and has the advantage of applying to k-dimensional manifolds in n-dimensional Riemann (or even Finsler) spaces.

12. Differentiability properties of geodesics

Our next subject concerns the properties of geodesics on convex hypersurfaces. We begin with differentiability questions which,

[2] The term "intrinsic area" is there used in a different sense: for each arcwise connected subset its intrinsic distances enter the evaluation of its Hausdorff measure, so that, for example, the k-dimensional measure of a nowhere differentiable arc in E^n is infinite for all k. Therefore this area is not monotone. Nevertheless it is very useful in all questions pertaining to general differential geometry, because the definition corresponds exactly to the procedure of classical differential geometry and the lack of monotoneity is outweighed by the guarantee that the concepts are intrinsic. Unfortunately the author suggested in [1] that this area might replace Lebesgue area, which in the meantime has proved to be the most important extrinsic area in E^n.

properly, belong to extrinsic geometry. Since these are local we may consider segments. The first results were obtained in B.F.[1] where it is proved that a segment issuing from, or passing through, a point b of a convex surface C where C is differentiable, has at b a tangent and an osculating plane which is normal to the tangent plane. Later Liberman [1] treated this question for a general point of a convex hypersurface with a particularly elegant method which we will now explain.

Consider a segment $T(b, q)$, $q \neq b$ on a complete convex hypersurface C. Take any line L through b which contains points of the

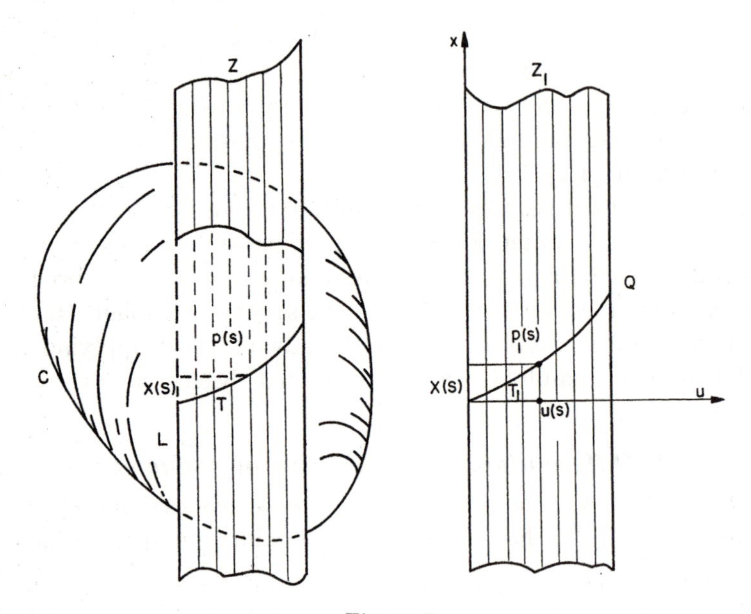

Figure 7

open convex set K bounded by C. For points x sufficiently close to b, say on the subsegment T of $T(b, q)$, the line parallel to L through x will also contain points of K. These lines form a cylinder Z, which can be developed on a plane Q because T is rectifiable. Denote the image in Q of a set or point in Z by the same symbol with the subscript 1. Then Z_1 is a strip between parallel lines, one of which is L_1. We claim that T_1 is a *convex curve*. More precisely, if D is that part of Z bounded by T which contains points of K, then D_1 is convex.

The proof is very simple: if the assertion were false then T_1 would contain points c_1 and d_1 such that $E_1 = E(c_1, d_1)$ lies, except for c_1, d_1 in $Z_1 - D_1$. The image of E_1 on Z is a curve E connecting the points c and d on T and lying outside K except for c and d. By (11.2) the length of E exceeds that of the subsegment $T(c, d)$ of T. Therefore E_1 would be longer than the arc from c_1 to d_1 of T_1.

Introduce rectangular coordinates u, x in Q with b_1 as origin such that L_1 is the x-axis with $x > 0$ in D_1 and $u > 0$ crossing Z_1. Then x is a convex function $f(u)$ of u. We consider x and u as functions $x(s)$, $u(s)$ of the arc length s on T_1 measured from b_1 to $p_1 = (u, x)$. The arc of T from b to p also has length s and $\pm x(s)$ is also the distance from b of the projection of p on L. We deduce from (11.8) that

$$\lim_{h \to 0} |h|^{-1} \{[x(s+h) - x(s)]^2 + [u(s+h) - u(s)]^2\}^{1/2} = 1.$$

Since $f(u)$ has everywhere one-sided derivatives $f'_r(u)$, $f'_l(u)$, we see from the last equation that

$$x'_r(s) = f'_r(u)\,[1 + (f'_r(u))^2]^{-1/2}, \quad x'_l(s) = f'_l(u)\,[1 + (f'_l(u))^2]^{1/2}$$

exist and are non-decreasing, because $f'_r(u)$ and $f'_l(u)$ have this property. Moreover, $x'_r(s) = x'_l(s)$ except for an at most denumerable set of values s, and $x'_r(s)$ has almost everywhere a second derivative.

We now take n independent lines $L^1, \ldots, L^n$ through b which contain points of K and consider L^i as an x_i-axis of a, not necessarily rectangular, coordinate system with $x_i > 0$ intersecting K. For a sufficiently small subsegment T of $T(b, q)$ a line parallel to L^i through points of T will intersect K for all i. Denote by $x_i(s)$ the abscissa of the projection of $p(s)$ on L^i. The $x_i(s)$ are the covariant components of the vector $\overrightarrow{bp}$ and $x_i(s)$ is exactly our previous function $x(s)$ with respect to the line L^i. The existence of derivatives of a vector is independent of the choice of the coordinate system and the type of component we use, hence we have proved:

(12.1) THEOREM. *Let* $\mathfrak{v}(s)$ *represent the vector leading from a fixed point z to the variable point* $p(s)$, $\alpha \leqq s \leqq \beta$, *of a segment G on a convex hypersurface C, where s is arc length on G. Then* $\mathfrak{v}(s)$ *has everywhere one-sided derivatives* $\mathfrak{v}'_r(s)$ *and* $\mathfrak{v}'_l(s)$ *with* $|\mathfrak{v}'_r(s)| = |\mathfrak{v}'_l(s)| = 1$. *Moreover,* $\mathfrak{v}'_r(s) = \mathfrak{v}'_l(s)$ *except for an at most denumerable set of values s and* $\mathfrak{v}(s)$ *has almost everywhere a second derivative.*

The convexity of the arc T_1 in Q above entails further:

(12.2) *Let* $T(b, q)$ *have the (right) tangent L at b. The limit H of any converging sequence of half planes* H_i *which are bounded by L and pass through a point* $q_i \in T(b, q)$ *with* $q_i \to b$ *forms an angle* $\leqq \pi/2$ *with any half plane bounded by L and containing a point of K.*

The simple proof is left to the reader.

This implies

(12.3) *If C is differentiable at b then any segment issuing from, or passing through, b possesses at b an osculating plane which is normal to the tangent plane of C at b.*

With the notation of (12.1) let $q = p(s)$, $\alpha < s < \beta$. If T_q is the tangent cone of C at q then the two semitangents R_1, R_2 of G at q are generators of T_q. If $p_i \in R_i$ and $|p_i - q| = 1$ then it follows from (11.4) that

$$\frac{(p_1q, T_q) + (qp_2, T_q)}{(p_1p_2, T_q)} = \frac{2}{(p_1p_2, T_q)} = \lim_{h\to 0+} \frac{2h}{(x(s-h)\,x(s+h), C)}.$$

Therefore the subarc from $p(s-h)$ to $p(s+h)$ of G can be a segment only if $E(p_1, q) \cup E(q, p_2)$ is a segment on T_q. If q is a *conical point* of C, i.e., if C possesses n independent supporting planes at q, then $E(p_1, q) \cup E(q, p_2)$ is never a segment on T_q, hence G cannot pass through a conical point. If T_q is a hyperplane then $E(p_1q) \cup E(q, p_2)$ must be $E(p_1, p_2)$ or G possesses a tangent at q. Thus

(12.4) *With the notations of* (12.1) *no point* $p(s)$ *with* $\alpha < s < \beta$ *is a conical point of* C; *and* $\mathfrak{v}(s)$ *possesses a derivative, or* G *a tangent, at every point* $p(s)$ *where* C *is differentiable.*

For $n > 3$ there are various cases between the two extremes where q is conical or T_q is a hyperplane, which could easily be discussed. For $n = 3$ there is only one more case where T_q consists of two non-coplanar half planes H_1, H_2 with a common boundary line L. If, in that case, q decomposes L into the rays L_1 and L_2 then $\angle(R_1, L_1) = \angle(R_2, L_2)$ for the two semitangents R_1, R_2 of G at q.

In spite of their surprising differentiability properties *segments can, on general convex surfaces, behave very erratically in other respects.* It is necessary to be aware of the possibilities in order to understand the difficulties that we will face when studying the intrinsic geometry of general convex surfaces.

As a first example we consider a *lense* Λ formed by a spherical cap smaller than a hemisphere and its image under reflection in the plane containing the boundary B of the cap. Clearly, *two distinct points of* B, *no matter how close, can be joined by two segments, and there is no segment issuing from a point* b *of* B *and tangent to* B *at* b. The same effect can be obtained on a differentiable surface Λ' obtained from Λ by smoothing it suitably along B; the normal curvature will be infinite in the direction normal to the curve corresponding to B.

For a second example take in the plane $x_2 = 0$ a closed convex curve D tangent at $b = (0, 0,1)$ to the circle $x_1^2 + x_3^2 = 1$ containing this circle, except for b, in its interior and with curvature 1 at b. On D select a sequence of points q_i tending to b but such that the tangent cones to the unit sphere Z from different q_i (taken up to their points of contact with Z) do not intersect. Then the convex closure of Z and the points q_i is a "Kappenkörper" C of Z. Since D has curvature 1 at b, this point is an Euler point, in fact the indicatrix is a circular disk with radius 1 and center b. By (12.4) no segment passes through any q_i, whence it easily follows from reasons of symmetry that there is a sequence of points r_i on the intersection of C with $x_2 = 0$ for which two segments $T(b, r_i)$

exist. We may even round off the apex of C_i such that we do not interfere with these segments (they remain segments on the new surface by (11.2)). Then we have *a surface where the indicatrix exists at every point and contains the point in the interior.* If the q_i approach b from both sides then *no segment will exist on C with either of the rays* $x_1 \geqq 0$, $x_3 = 1$, $x_2 = 0$ or $x_1 \leqq 0$, $x_3 = 1$, $x_2 = 0$ as tangent. If $x_1 > 0$ for all q_i then *a segment* $T(y, b)$, $y_2 = 0$, $y_1 < 0$, $y \in Z$ *cannot be prolonged beyond* y, and stay a segment no matter how close y is to b.

Finally, it is well known that there are convex surfaces with an everywhere dense set of conical points (see, for example, A., pp. 59, 60). On such a surface the tangent cone T_q at an arbitrary point q contains a dense set of generators which are not semitangents of segments issuing from q.

However, it is shown in A., pp. 213, 214 that at *a given point q of an arbitrary convex surface almost all generators* (in an obvious sense) *of the tangent cone* T_q *at q are semitangents of segments.*

Various questions now arise. The first concerns weaker conditions than the classical which exclude these occurrences. The following is proved in B. F. [2, I] by approximation with analytic surfaces:

(12.5) THEOREM. *Let the point p of the differentiable convex surface C have a neighborhood in which the (one-sided) upper normal curvatures are bounded.*[3] *Then a disk* $D : px < \varrho$, $\varrho > 0$ *on C exists with the properties: the segment* $T(a, b)$ *is unique for* a, $b \in D$. *For a given line element at p there is exactly one segment with mid-point p, length* 2ϱ *and tangent to this line element. These segments cover D (except for p) simply.*

If the one-sided normal curvatures of a complete convex surface are smaller than $1/R$ *then* $T(a, b)$ *is unique for* $ab < \pi R$.

The next question concerns the differentiability properties of segments on non-convex surfaces. Nothing seems to be known beyond the classical case for higher dimensions. For surfaces no theorem is available which comprises (12.1) or (12.3) as special

[3] Equivalently: the lower indicatrix at any point in the neighborhood contains a circular disk with a fixed radius about the point.

case;[4] the strongest result is contained in Busemann and Mayer [1] which deals with general variational problems $\int F(u, \dot{u})dt$ in parametric form:

Let $F(u_1, u_2, \xi_1, \xi_2) = F(u, \xi)$ *be continuous in a certain u-domain and for all* ξ, *and satisfy the usual conditions:* $F(u, \xi) > 0$ *for* $\xi \neq 0$, $F(u, k\xi) = kF(u, \xi)$ *for* $k \geqq 0$ *and the (Legendre) condition: the curve* $F(u_0, \xi) = 1$ *in the* ξ*-plane is strictly convex for each* u_0. *If*

$$|F(\bar{u}, \xi) - F(u, \xi)| \leqq C\,|\bar{u} - u| \cdot |\xi|, \tag{12.6}$$

then every Hilbert arc (segment in our language) is of class C^1.

But a Hilbert arc need not possess second derivatives, and a given line element may be on infinitely many Hilbert arcs which do not coincide locally, or on none. Thus, we have a similar phenomenon as for convex surfaces: the continuous differentiability of segments follows from rather weak assumptions, uniqueness and existence (of segments through given line elements) do not. The condition (12.6) is satisfied for a Riemannian line element

$$F(u, \xi) = [E(u)\xi_1^2 + 2F(u)\xi_1\xi_2 + G(u)\xi_2^2]^{1/2},$$

if E, F, G satisfy Lipschitz conditions $|E(\bar{u}) - E(u)| < c|\bar{u} - u|, \ldots$. Hartman and Wintner [1] furnish very instructive examples which show that *even if* E, F, G *are of class* C^1 *and if their first partial derivatives satisfy Hölder conditions with an exponent less than* 1 *the above negative statements and various others remain true.* The geodesics need not be the solutions of the (now everywhere defined) differential equations $u''_i + \Gamma_{jk}^{i} u'_j u'_k = 0$. However (*op. cit.*), when the first partial derivative of E, F, G satisfy Lipschitz conditions (Hölder conditions with the exponent 1) then the situation is very much as in the classical case.

According to a surprising result of Hartman [1], the solutions

[4] Added during proof: In view of two recent publications these statements must be modified. Reshetnyak [2] extends (12.1) roughly to those hypersurfaces which, for some fixed positive δ, possess everywhere supporting spheres of radius δ. Siegel [1] proves that Hilbert arcs are of class C^1 for variational problems belonging to an integrand $F(u_1, \ldots, u_n, \xi_1, \ldots, \xi_n)$ satisfying the analogous conditions to those below, but assuming instead of (12.6) that $\partial F/\partial \xi_i$ exists for $\xi \neq 0$ and that $\partial F/\partial u_i$ is continuous.

of $u''_i + \Gamma_{jk}^i u'_j u'_k = 0$ have, without Hölder conditions on E, F, G, the uniqueness properties of the first part of (12.5) when E, F, G are derived from a surface $\mathfrak{x}(u, v)$ of class C^2 in E^3 with $\mathfrak{x}_u \times \mathfrak{x}_v \neq 0$, i.e., $E = \mathfrak{x}_u^2$, $F = \mathfrak{x}_u \cdot \mathfrak{x}_v$, $G = \mathfrak{x}_v^2$.

13. Angles. The convexity condition

Our principal aim is the characterization of those abstract two-dimensional metrics which can be realized as intrinsic metrics on convex surfaces in E^3. The corresponding problem for higher dimensions has not been solved and has, at any rate, a completely different character, because hypersurfaces in E^n, $n > 3$, are, in general, already locally rigid. Therefore we consider from now on *only surfaces in E^3 and two-dimensional abstract manifolds.*

Most of the material in Sections 13–20 is in Alexandrov's book. Here we merely report partly without proofs, with the intention of providing the reader with a clear understanding of the guiding principles: the main results are very beautiful because they supply, without unnatural restrictions, easily grasped answers to deep problems. But the proofs, which are at present available, are among the longest in the mathematical literature.

The basic concept in Alexandrov's theory is the *angle between segments*. Two proper segments S' and S'' issuing from the same point p of a convex surface C have semitangents R' and R'' at p which are generators of the tangent cone T_p of C at p and divide T_p into two sectors. The angles between R' and R'' on T_p are the angles of these sectors after they have been unfolded on the plane. The sum of these angles is the complete angle of C at p. It is 2π except at the conical points, where it is smaller.

The smaller of the two angles between R' and R'' on T_p is called the *angle between S' and S'' at p.* (We do not exclude the case that $R' = R''$, and hence this angle is 0.)

Because a segment S does not pass through a conical point each of the angles between the two segments S' and S'' into which an interior point of p of S decomposes S equals π. If S^0 is any segment issuing from p then the sum of the angles between S^0 and S' and between S^0 and S'' equals π.

This definition of angle is extrinsic, but it can be given an intrinsic form by means of the following fact (A., pp. 172, 173), which follows very easily from (11.4).

(13.1) *If* $x'_\nu \in S'$, $x''_\nu \in S''$, $x'_\nu \to p$, $x''_\nu \to p$ *and*

$$0 < \delta < x'_\nu p / x''_\nu p < \delta^{-1},$$

then the angle α *between* S' *and* S' *at* p *is determined by*

$$\cos \alpha = \lim_{\nu \to \infty} (px_\nu'^2 + px_\nu''^2 - x'_\nu x_\nu''^2)(2px'_\nu \cdot px''_\nu)^{-1}.$$

We will see presently that the restriction on $x'_\nu p / x''_\nu p$ is superfluous; but (11.4) alone will not give this result.

For an intrinsic characterization of the metric on general convex surfaces with angle as basic concept, we need a definition of angle in general spaces. Guided by (13.1) we denote for three points p, x, y, $p \neq x$, $p \neq y$ in a metric space by $p(x, y)$ the angle defined by

$$\cos p(x, y) = \frac{px^2 + py^2 - xy^2}{2px \cdot py}, \quad 0 \leqq p(x, y) \leqq \pi.$$

It is the angle opposite the side $x'y'$ in a euclidean triangle $p'x'y'$ isometric to pxy, i.e., $|p' - x'| = px, \ldots$.

Consider two Jordan arcs $x(t)$, $y(t)$, $0 \leqq t \leqq 1$, with $x(0) = y(0) = p$ in a metric space. We say that *the angle* α *between these curves at* p *exists if the double limit*

$$\alpha = \lim_{s \to 0,\, t \to 0} p(x(s), y(t)) \tag{13.2}$$

exists.

In order to become familiar with the angle (13.5) we prove two simple general facts:

(13.3) *Triangle Inequality. In a metric space let*

$$x_i(t),\ i = 1, 2, 3,\ 0 \leqq t \leqq a_i,\ a_i > 0,\ x_i(0) = p$$

be three Jordian curves for which the angles α_{ij} *between* $x_i(t)$ *and* $x_j(t)$ *at* p *exist. Then*

$$\alpha_{13} \leqq \alpha_{12} + \alpha_{23}.$$

If t_1 and t_3 are such that with $x_i(t_i) = x_i$ the relations

$$0 < px_1 = px_3 < px(a_2)$$

hold, construct in E^2 a triangle qy_1, y_3 with

$$|q - y_1| = |q - y_3| = px_1 = px_3, \quad |y_1 - y_3| = x_1x_3$$

so that $q(y_1, y_3) = p(x_1, x_3)$. We show that for a suitable t which tends to 0 with t_1 and t_3

$$q(y_1, y_3) = p(x_1, x_3) \leqq p(x_1, x_2(t)) + p(x_2(t), x_3).$$

This is obvious for $x_1 = x_3$. We assume therefore $x_1 \neq x_3$ hence $y_1 \neq y_3$ and denote by $h < qy_i$ the distance of q from $L(y_1, y_3)$. Let t' be the last value of t for which $px_2(t) = h$ and t'' the first value following t' for which $px_2(t) = px_i$. As t traverses the interval $[t', t'']$ the function $x_1x_2(t)[x_1x_2(t) + x_2(t)x_3]^{-1}$ varies continuously and takes values in $[0,1]$. For each $t \in [t', t'']$ we take the two (one if $qx_2(t) = h$) points v_i on $E(y_1, y_2)$, $|y_1 - v_1| \leqq |y_1 - v_2|$, for which $|q - v_i| = px_2(t)$. Then $|y_1 - v_1|/|y_1 - y_3|$ traverses $[0, \frac{1}{2}]$ whereas $|y_1 - v_2|/|y_1 - y_3|$ traverses $[\frac{1}{2}, 1]$. Therefore a t in $[t', t'']$ exists such that with $w = v_1$ or $w = v_2$

$$\frac{|y_1 - w|}{|y_1 - y_3|} = \frac{x_1x_2(t)}{x_1x_2(t) + x_2(t)x_3}, \text{ hence } \frac{|w - y_2|}{|y_1 - y_3|} = \frac{x_2(t)x_3}{x_1x_2(t) + x_2(t)x_3}.$$

Since

$$|y_1 - y_3| = x_1x_3 \leqq x_1x_2(t) + x_2(t)x_3,$$

we conclude $|y_1 - w| \leqq x_1x_2(t)$ and $|w - y_2| \leqq x_2(t)x_3$. Therefore $|q - w| = px_2(t)$ and $px_i = qy_i$ entail

$$p(x_1, x_2(t)) + p(x_2(t), x_3) \geqq q(y_1, w) + q(y_2, w) = q(y_1, y_3).$$

Without the assumption that the angles α_{ij} exist (13.3) is proved in A., pp. **140–143** for the corresponding superior limits.

The triangle inequalities for distances and for angles yield

(13.4) *Let $x_i(t)$, $i = 1, \ldots, n$ be Jordan curves in a metric space issuing from p, ($x_i(0) = p$), and let a segment connecting points $x_1 = x_1(t'_\nu)$, $t'_\nu \to 0$, and $x_n = x_n(t''_\nu)$, $t''_\nu \to 0$, exist which intersects* *the curves* $x_2(t), \ldots, x_{n-1}(t)$ *in this order in points* $x_2, \ldots, x_{n-1}$ *different from* p. *If the angle* $\alpha_{i,i+1}$ *between* $x_i(t)$ *and* $x_{i+1}(t)$ *exists, then the angle* $\alpha_{1,n}$ *between* $x_1(t)$ *and* $x_n(t)$ *exists and*

$$\alpha_{1n} = \alpha_{12} + \alpha_{23} + \ldots + \alpha_{n-1,n}.$$

For since $y_1p + py_n \geqq y_1y_n$, we can construct points q, y_i, $i = 1, \ldots, n$, in E^2 such that the rays $R(q, y_1)$ and $R(q, y_n)$ bound a convex or a straight angle containing the points $y_2, \ldots, y_n$ and the triples q, y_i, y_{i+1} are isometric to p, x_i, x_{i+1}. Then, since the x_i lie in the order $x_1, x_2, \ldots, x_n$ on a segment

$$|y_1 - y_n| \leqq \sum |y_i - y_{i+1}| = \sum x_ix_{i+1} = x_1x_n,$$

hence

$$\sum p(x_i, x_{i+1}) = \sum q(y_i, y_{i+1}) = q(y_1, y_n) \leqq p(x_1, x_n)$$

which together with the proof of (13.3) yields the assertion.

On a convex surface we know so far the existence of angles between segments only under a restriction, see (13.1), which we now eliminate. It is easily seen that every point of a smooth surface with non-negative curvature has a neighborhood which satisfies the

(13.5) *Convexity Condition. If* $p \neq x$ *and* $(py'y)$ *then* $p(x, y) \leqq p(x, y')$.

Applying the convexity condition twice yields:

$$(px'x) \text{ and } (py'y) \text{ entail } p(x, y) \leqq p(x', y').$$

It is a fundamental fact that the convexity condition is not only fulfilled locally:

(13.6) THEOREM. *A complete convex surface*[5] *satisfies the convexity condition in the large.*

The proof of the theorem is far from trivial. It is proved for polyhedra first, A., pp. 115–124, and then for general convex surfaces by approximation with polyhedra, A., pp. 125–130. One of the principal difficulties is that the convexity condition deals with the distances xy, and that the segments $T(x, y)$ which do not depend continuously on x and y, enter the proof.

[5] We now always include doubly covered plane convex domains.

The great importance of the convexity condition lies in the fact that it *characterizes* the intrinsic metrics of convex surfaces. We will see that any intrinsic metric on S^2, or any finitely compact intrinsic metric on E^2, which satisfies the convexity condition can be realized as the intrinsic metric of a convex surface in E^3.

Because (13.6) requires a long proof, the convexity condition is too recondite to be entirely satisfactory as a sufficient condition. For this reason, Alexandrov gives an equivalent condition in terms of the excess of geodesic triangles with which we are essentially familiar, see Section 18. (13.6) implies, of course:

(13.7) *A non-complete convex surface satisfies the convexity condition locally.*

We now consider a (two-dimensional) manifold M with an intrinsic metric xy satisfying the convexity condition. Consider two segments S, T on M issuing from p and represented in terms of arc length by

(13.8) $$S: \ x(s),\ 0 \leqq s \leqq \alpha, \qquad T: \ y(t),\ 0 \leqq t \leqq \beta; \\ x(0) = y(0) = p, \quad \alpha \geqq \beta > 0.$$

Put $p(x(s'), y(t)) = p(s, t)$. Then the angle

(13.9) $$\delta = \lim_{t \to 0,\, s \to 0} p(s, t) \text{ exists and } \delta \geqq p(s, t).$$

Moreover, $\delta = 0$ *only when* $S \subset T$.

For, the convexity condition yields

$$p(s, t) \leqq p(s', t') \text{ if } s' \leqq s \text{ and } t' \leqq t.$$

If $S \not\subset T$ then $x(t) \neq y(t)$ for a suitable $t \leqq \beta$, hence $p(t, t) > 0$ and $\delta > 0$.

(13.10) *Inclusion Property. With the notation* (13.8), *if* $x(t) = y(t)$ *for* $0 \leqq t \leqq \beta' < \beta$, $\beta' > 0$, *then* $S \subset T$ *or* $x(t) = y(t)$ *for* $0 \leqq t \leqq \beta$.

For $p(t, t)$ is non-increasing and 0 for $0 < t \leqq \beta'$. An immediate corollary of (13.10) is

(13.11) *If* (abc) *and segments* $T(a, b)$, $T(b, c)$ *exist, then they are unique.*

If T' and T'' are two segments from b to c and T_0 is any $T(a, b)$

then $T_0 \cup T'$ and $T_0 \cup T''$ are by (10.10) segments $T(a, c)$ which coincide from a to b, therefore they coincide from a to c.

The inclusion property alone has other important implications which will be discussed later. The convexity condition entails:

(13.12) *With the notations* (13.8), *if* $s_\nu \to 0$, $t_\nu \to t_0$, $0 < t_0 < \beta$, *and a suitable segment* $T(x(s_\nu), y(t_\nu))$ *tends to the (unique) segment* $T(p, y(t_0))$ *then*

$$\lim_{\nu \to \infty} p(s_\nu, t_\nu) = \delta = \lim_{s \to 0,\, t \to 0} p(s, t).$$

This fact is described by Alexandrov as the existence of the angle between S and T at p in "*the strong sense*," Its significance will appear later. The proof is lengthy, A., pp. 134–138.

(13.13) *Let* T_i *be a segment connecting* a_{i-1} to a_{i+1}, $i = 1, 2, 3$, $a_{\pm 3+k} = a_k$ *and* α_{i+1} *the angle between* T_{i-1} *and* T_i at a_{i+1}. *Then* α_{i+1} *is at least as large as the angle* α'_{i+1} *at* a'_{i+1} *in the euclidean triangle* $a'_1 a'_2 a'_3$ *with* $|a'_{i-1} - a'_i| = a_{i-1} a_i$.

Briefly: the angles in a geodesic triangle equal at least the corresponding angles in an isometric euclidean triangle; consequently, *the sum of the angles is at least* π (see Figure 9). This fact is well known for small triangles on smooth surfaces with non-negative curvature.

The proof is obvious: If T_1 and T_2, say, are represented by S and T as in (13.8) then $\alpha'_3 = p(\alpha, \beta) \leqq \delta = \alpha_3$.

We precede the next statement by an example. Let C be the boundary of a cube; p a vertex of the cube and a, b, c the end points of the edges issuing from p. If p_ν lies on the edge $E(p, c)$ then the oriented segments $E^+(p_\nu, a_\nu)$ and $E^+(p_\nu, b_\nu)$ parallel to, and of the same length as, $E^+(p, a)$, $E^+(p, b)$ lie on C. These euclidean segments $E(p, a), \ldots$ are also segments $T(p, a), \ldots$ for the intrinsic metric on C. With respect to this metric the angle between $T(p_\nu, a_\nu)$ and $T(p_\nu, b_\nu)$ is π, whereas the angle between $T(p, a)$ and $T(p, b)$ is $\pi/2$. Thus for $p_\nu \to p$ the former angle does not tend to the latter, although $T(p_\nu, a_\nu) \to T(p, b)$ and $T(p_\nu, b_\nu) \to T(p, b)$.

This example shows that the following theorem cannot be improved:

(13.14) *Lower Semicontinuity of Angles: Let the closed convex surface C_i tend to C and let S_i, T_i be segments on C_i issuing from p_i with the property that $p_i \to p \in C$ and S_i, T_i tend to proper segments S, T on C. If δ_i, δ are the angles between S_i and T_i on C_i and between S and T on C then lim inf $\delta_i \geqq \delta$.*

In the proof we use the notations of (13.8) with the subscript i for S_i and T_i. There is a $\beta > 0$ such that

$$\delta < p(s, t) + \epsilon \text{ for } 0 < s,\ t < \beta.$$

Let $x_i(s_i) \to x(s)$, $y_i(t_i) \to y(t)$, $0 < s$, $t < \beta$. Then (11.3) yields

$$s_i \to s,\ t_i \to t \text{ and } (x_i(s_i)y_i(t_i), C_i) \to (x(s)y(t), C).$$

Now (13.9) shows

$$\delta_i \geqq p_i(s_i, t_i) \to p(s, t) > \delta - \epsilon.$$

Consider *a two-dimensional manifold M with an intrinsic metric which satisfies the inclusion property and for which the angle* (13.2) *for any two segments S, T issuing from a point p exists.* These segments divide a suitable neighborhood of p which is homeomorphic to E^2 into two sectors V, V' homeomorphic to closed angular domains in E^2. Because of the inclusion property it is meaningful to speak of segments $S_1 = S, S_2, \ldots, S_n = T$ in V and following each other in this circular order (we do not assume that $S_2, \ldots, S_{n-1}$ exist for $n > 2$). If α_{ij} is the angle between S_i and S_j then (13.3) gives

$$\alpha_{1n} \leqq \alpha_{12} + \alpha_{23} + \ldots + \alpha_{n,n-1}.$$

As the angle $\alpha(V)$ of the sector V (sector angle between S and T) we define the least upper bound of

$$\sum_{i=1}^{n-1} \alpha_{i,i+1}$$

for all choices of $S_1, \ldots, S_n$ in V (n is variable). Then

$$\alpha(V) \geqq \alpha_{1n}$$

and the equality sign holds by (13.4) if there are points $x \in S$ and $y \in T$ arbitrarily close to, but different from, p such that a $T(x, y) \subset V$ exists.

If the segment S' begins at p and lies in V, then it divides V into two sectors V_1, V_2. It follows from (13.3) that $\alpha(V_1) + \alpha(V_2) \geqq \alpha(V)$.

The sum $\alpha(V) + \alpha(V')$ is the *complete angle* of M at p. Because of the additivity of sector angles it does not depend on the choice of S and T, but may be infinite without further assumptions on the metric. We conclude from (13.1), (13.6), (13.9):

(13.15) *The angle between two segments S', S'' issuing from a point p of a convex surface C equals the sector angle $\alpha(V)$ of one of the sectors determined by S', S'' and the angle of the corresponding sector on the tangent cone T_p of C at p. The complete angle of C at p is at most 2π.*

14. Triangulations. Intrinsic curvature

The irregular behavior of segments on general convex surfaces, see Section 12, shows that the existence of triangulations of a general convex surface into small geodesic triangles is not obvious. For the realization theorems of the next chapter triangulations for even more general two-dimensional manifolds are needed. Various results are known, it suffices for us to construct *triangulations on the basis of the inclusion property.* The construction, if carried out in every detail, becomes lengthy, A., pp. 87–105. The basic ideas are, however, very simple and we will describe them briefly.

We assume throughout that we deal with a two-dimensional manifold M with an intrinsic metric and the inclusion property. A *polygon* is a curve of the form $\Sigma_{i=1}^{n} T(a_i, a_{i+1})$; it is closed when $a_{n+1} = a_1$ and *simple* when it is a Jordan curve. A compact set on M with interior points is a *polygonal domain* if its boundary consists of a finite number of disjoint simple closed polygons.

In particular, a *triangle* is a set homeomorphic to a closed disk with a boundary of the form $T(a_1, a_2) \cup T(a_2, a_3) \cup T(a_3, a_1)$. Since segments are shortest connections, $a_1a_2 \geqq a_2a_3 + a_3a_1, \ldots$.

The equality sign may hold. For example, if T_1 and T_2 are two distinct segments from a_1 to a_2 on a closed convex surface C and a_3 is an interior point of T_1 then the (by (13.11) unique) segments $T(a_1, a_3)$ and $T(a_3, a_2)$ bound with T_2 two triangles on C in which $a_1a_2 = a_2a_3 + a_3a_1$. A triangle is called *proper* if the sum of any two sides is greater than the third.

A subset of M which contains with any two points x, y at least one segment $T(x, y)$ is called *convex.* [6] The preceding example shows that a triangle need not be convex. A *convex domain* is a compact convex set with interior points whose boundary consists of a finite number of disjoint closed Jordan curves. For both polygonal and for convex domains we admit the possibility that the boundary is empty, e.g., if M is homeomorphic to S^2 then it is a polygonal convex domain.

For the existence of triangulations it is first shown, A., pp. 87–94, that a *given point p is interior point of a polygonal convex domain homeomorphic to a disk and with arbitrarily small diamater.* It suffices to find such a domain in a given neighborhood $U(p, 5\varrho)$, $\varrho > 0$. We choose ϱ so small that $T(x, y)$ exists for $x, y \in U(p, \varrho)$, see (10.11), and draw a closed Jordan curve G in $U(p, \varrho)$ which contains p in its interior. Take two distinct points u, v on G and a topological mapping $x \to x'$ of one, A, of the two arcs of G from u to v on the other, A', such that $u \to u$ and $v \to v$. A segment $T(x, x')$ lies by (10.8) in $U(p, 2\varrho)$ and close to u (v) when x lies close to u (v). There are two possibilities:

1) *There is a $T(x, x')$ which passes through p.* We then choose a z on $T(p, x')$ close to p and points z_1, z_2 on different sides of $T(x, x')$ close to z. The inclusion property guarantees that $T(p, z_i)$ has no other common points with $T(x, x')$ than x. If the z_i are sufficiently close to z then $T(x, z_1) \cup T(z_1, z_2) \cup T(z_2, x)$ bounds a, not necessarily convex, triangle containing p in its interior which will lie in $U(p, 2\varrho)$.

[6] A convex set on a convex surface may be quite pathological: A set obtained from a closed convex surface by omitting some or all points of the possibly everywhere dense set of conical points contains, because of (12.4), with any two points *every* segment $T(x, y)$.

2) *If no* $T(x, x')$ *passes through* p we observe that the curve consisting of $T(x, x')$ and the arcs of A from u to x and of A' from u to x' can be contracted to u (in a neighborhood of p) without passing through p when x is close to u, and that this is not so when x is close to v. This yields easily, for a suitable position of x, two segments $T(x, x')$ such that the domain in $U(p, 5\varrho)$ bounded by them contains p in its interior. As in the above example we may consider this set as a (non-proper) triangle.

To obtain a *convex* polygonal domain we consider a shortest closed curve in $U(p, 5\varrho)$ containing the triangle constructed under 1) or 2). It is easily seen that the minimum is attained in $U(p, 5\varrho)$, that a minimizing curve D bounds a set homeomorphic to a disk, which is convex because, otherwise, D could be shortened. For the same reason D must be a segment in the neighborhood of every point which is not a vertex of the triangle.

Thus D consists of at most three geodesic arcs joined at vertices of the triangle. These arcs can be divided into segments. The domain bounded by D and containing p satisfies our requirements.

For the second part of the existence proof for triangulations, A., pp. 95–105, we observe first:

(14.1) *If* x *is an interior point of a convex domain* B *and* y *is any point of* B, *then every segment* $T(x, y)$ *lies in* B.

For, if there were a point q on $T(x, y)$ outside B, choose q' in $T(x, y) \cap B$ between x and q. Then $T(q', y)$ is by (13.11) unique, but contains $q \notin B$.

It follows that *the intersection of two convex domains* B_1, B_2 *is convex when it contains interior points.* For if $x, y \in B_1 \cap B_2$ and x is an interior point of $B_1 \cap B_2$, then every $T(x, y) \subset B_1 \cap B_2$ by the preceding remark. If x lies on the boundary of $B_1 \cap B_2$ it can be approximated by interior points x_n of $B_1 \cap B_2$. Then $T(x_n, y) \subset B_1 \cap B_2$ and, because of (10.5), a subsequence of $\{T(x_n, y)\}$ will converge to a segment $T(x, y) \subset B_1 \cap B_2$.

Hence, the intersection of two convex polygonal domains P_1, P_2 is again such a domain, provided it contains interior points. For we know that $P_1 \cap P_2$ is a convex domain. That its boundary which consists of parts of the boundaries of P_1 and P_2 is the union of a finite number of segments follows from the inclusion property.

These facts lead through simple, but lengthy, arguments to the result that the union of a finite number of convex polygonal domains which are homeomorphic to a disk can be represented as the union of a finite number of polygonal domains homeomorphic to disks without common interior points, each of which lies in one of the given domains.

It is easily seen that a convex polygonal domain which is homeomorphic to a disk can be triangulated into proper convex triangles, essentially by drawing suitable diagonals. A given polygonal domain can by the first part be covered by a finite number of convex polygonal regions homeomorphic to disks and with arbitrarily small diameter. Using the second part we obtain the following result:

(14.2) THEOREM. *A polygonal domain (on a two-dimensional manifold with an intrinsic metric and the inclusion property) can be triangulated into proper triangles with arbitrarily small diameter. If the given domain is convex then the triangles can be chosen convex.*

It is an open question whether the triangles can be made convex also when the given domain is not convex. It is not hard to see that the answer is affirmative on convex surfaces, because the complete angle at any point is at most 2π.

We now assume in addition to the inclusion property that *the angle* (13.2) *between segments exists, that the complete angle at any point is finite,*[7] *and that an interior point of a segment divides the segment into two segments which form two sector angles with measure* π. A convex surface has all these properties, see last section.

Consider a polygonal domain P with boundary B. If two segments S', S'' on B meet at the point q of B then one, V, of the two sectors determined by q lies in P. The angle of P at q, or of B in P at q, is the sector angle $\alpha(V)$. By hypothesis $\alpha(V) = \pi$ when q is an interior point of a segment on B. In particular, a

[7] It suffices that it be finite at those points which are vertices of the triangles occurring in the triangulations. This remark is used in Section 18.

triangle T has three angles α_1, α_2, α_3 corresponding to its three vertices. We put

$$\epsilon(T) = \alpha_1 + \alpha_2 + \alpha_3 - \pi$$

and call it the *excess of* (*the interior of*) T. For a point p of M we denote by $\epsilon(p)$ the complete angle of M at p minus 2π.

Consider a polygonal domain P on M and denote by $\alpha_1, \ldots, \alpha_n$ those angles in P at points of the boundary B of P which are different from π. If P is triangulated into the triangles $T_1, \ldots, T_t$ with vertices $p_1, \ldots, p_\nu$ *inside* D, then

$$(14.3) \qquad \sum_{i=1}^{t} \epsilon\,(T_i) + \sum_{j=1}^{\nu} \epsilon\,(p_j) = 2\pi\chi(P) - \sum_{k=i}^{r} (\pi - \alpha_k)$$

where $\chi(P)$ is the *Euler characteristic* of P. This formula is well known when $\epsilon(p_i) = 0$ and it is very easily seen, see A., pp. 181–183, that the standard proof yields, under the present hypothesis, the corrective term $\Sigma\,\epsilon(p_j)$.

The right side of (14.3) shows that the left side is independent of the triangulation of P; we call it the *excess* $\epsilon(P)$ *of* P, or better, of the interior of P.

A *primitive* set on M is a point, a segment without its end points, or a triangle without its boundary. An elementary set is the union of a finite number of disjoint primitive sets. If W is an elementary set consisting of the points $p_1, \ldots, p_\nu$, certain open segments, and the triangles $T_1, \ldots, T_t$ then

$$(14.4) \qquad \epsilon(W) = \sum_{i=1}^{t} \epsilon(T_i) + \sum_{j=1}^{\nu} \epsilon(p_j)$$

is by definition, the *excess* of W. Thus if p is triangulated as above, then the interior of P is an elementary set consisting of the open triangles $T_1, \ldots, T_t$, their open sides and their vertices not on B. For this reason it is preferable to think of $\epsilon(P)$ *as the excess of the interior of* P.

If an elementary set W is represented in two different ways as a union of disjoint primitive sets then the evaluation of $\epsilon(W)$ by (14.4) leads, because of (14.3) and the inclusion property, to the same result, see A., pp. 184, 185. Since the union of two disjoint elementary sets is an elementary set we may say that the excess is an additive set function on the elementary sets.

(14.5) *On a convex surface C the excess $\epsilon(W)$ is non-negative.*

This follows from (13.13) and (13.15). We remember that (13.13) was deduced from the convexity condition which holds in the large only on complete surfaces. For non-complete surfaces we must therefore use triangles which lie in parts of C which satisfy the convexity condition.

On smooth surfaces there is a well-known connection between excess and Gauss curvature. An extension of this relation to general convex surfaces must evidently rest on the relation $\epsilon(W) = \nu(W)$, for primitive sets, where ν has the same meaning as in Section 4.

(14.6) *If T is an open triangle on a convex surface C then $\epsilon(T) = \nu(T)$.*

If S is an open segment on C then $\nu(S) = 0$.

For any point p of the surface $\epsilon(p) = \nu(p)$.

The proof in A., pp. 200–207 proceeds by approximation with polyhedra, for which these facts are easily established; in particular the second fact is obvious because a segment does not pass through a vertex of a polyhedron.

Since both the excess ϵ and the curvature ν are additive on the elementary sets it follows from (14.6) that

$$\epsilon(W) = \nu(W) \textit{ for all elementary sets } W.$$

We know from Section 4 that $\nu(M)$ can be extended to a non-negative, completely additive Set function on the Borel sets of C; moreover, $\nu(W)$, hence $\epsilon(W)$, is bounded, namely at most 4π. Well-known theorems on set functions, see Hahn-Rosenthal [1], entail that the extension of $\epsilon(W)$ to all Borel sets is unique, hence $\epsilon(M) = \nu(M)$ *for all Borel sets M of C.* In this definition of $\epsilon(M)$ we use $\nu(M)$, hence extrinsic properties of C. This may be rectified by observing that $\nu(M')$, hence $\epsilon(M')$, is for any closed bounded set M' the greatest lower bound of $\nu(W) = \epsilon(W)$ for all elementary sets $W \supset M'$. For an arbitrary Borel set M we obtain $\nu(M)$, hence $\epsilon(M)$ as the least upper bound of $\nu(M') = \epsilon(M')$, where M' traverses all closed bounded sets $M' \subset M$. The relation $\epsilon(M) = \nu(M)$ is, of course, a generalization of Gauss' theorem on the intrinsicness of his curvature.

(14.7) THEOREM OF GAUSS. *For any Borel set M on a convex surface the excess $\epsilon(M)$ equals the area $\nu(M)$ of the spherical image.*

We may, and will henceforth, use $\nu(M)$ both for the extrinsic and the intrinsic curvature or excess.

This approach to curvature led Alexandrov to an analogous approach to intrinsic area: A polygonal domain P on a convex surface can be triangulated into proper triangles $T_1^d, \ldots, T_{t_d}^d$ with diameters less than a given number d. For each triangle T_i^d take the area $A_e(T_i^d)$ of the euclidean triangle with the same sides, in other words, use Hero's formula to evaluate $A_e(T_i^d)$, and form $\Sigma A_e(T_i^d)$. Then $\Sigma A_e(T_i^d)$ tends for $d \to 0$ to a limit independent of the choice of the triangulations, which is called the area $A(P)$ of P. The proof yields the interesting estimate

$$0 < A(P) - \sum_{i=1}^{t_d} A_e(T_i^d) \leqq (1/2)\,\nu(P) d^2.$$

The area $A(P)$ coincides with the extrinsic area $A(P)$ used by us in Chapter I, and hence with the intrinsic area $|P|_2$ in terms of Hausdorff measure defined in Section 11. This extends to all Borel sets on the surface. These facts are found in A., pp. 391–407.

15. The Gauss-Bonnet theorem. Quasigeodesics

On a convex surface, (14.3) becomes, through (14.7), the Gauss-Bonnet theorem for polygonal domains. Our next aim is the extension of this result to more general domains, which requires an analogue to geodesic curvature or rather, its integral over a curve.

The concept of direction of a curve is needed. The definition (13.2) of angle can be applied to the case where $x(t) \equiv y(t)$. We say the arc $x(t)$ has at $x(0)$ *a definite* direction if the angle of $x(t)$ with itself exists; of course, the angle then equals 0. This definition will seem artificial at first sight, but the fact that a given line element may not lie on any segment precludes a direct approach by means of geodesic tangents. However, there is the following connection between angles formed by segments and by curves, A., p. 344:

(15.1) *The Jordan arcs $x(t)$, $y(t)$, $x(0) = y(0) = p$ form at p the angle α if, and only if, the angle $\alpha(s, t)$ between segments $T(p, x(s))$ and $T(p, y(t))$ tends to α for $s \to 0$, $t \to 0$.*

The proof is simple: if $\nu(s, t)$ is the curvature of the interior of the (small) triangle bounded by $T(p, x(s)) \cup T(x(s)), y(t)) \cup T(y(t), p)$ then $\nu(s, t) \to 0$, because ν is completely additive and the intersection of the open triangles is empty. If follows from (13.13), (14.5) and the fact that the excess of a euclidean triangle is 0 that

$$0 \leqq \alpha(s, t) - p(x(s), y(t)) \leqq \nu(s, t),$$

which implies the assertion.

In particular, the curve $x(t)$ has at $p = x(0)$ a definite direction if, and only if, the angle between $T(p, x(s))$ and $T(p, x(t))$ tends to 0 for $s \to 0$ and $t \to 0$.

It is more involved to prove, (A., pp. 345–347):

(15.2) *The angle between two Jordan arcs issuing from a point p of a convex surface exists, if and only if, each of the arcs has at p a definite direction.*

Therefore we may say that two curves issuing from the same point p have the same direction if their angle at p exists and equals 0. On the other hand, a Jordan arc $x(t)$ issuing from the point p of a convex surface C has at p a definite direction if, and only if, it has at p a semitangent, that is, if $\lim_{t\to 0} R(p, x(t))$ exists, A., pp. 348, 349. The angle between two Jordan arcs issuing from p, if it exists, equals therefore the angle measured on the tangent cone T_p of C at p. Thus there is a one-to-one correspondence between the directions on C at p and the generators of T_p: all curves with a given direction at p have the same generator of T_p as semitangent.

We now turn to geodesic curvature and consider first a geodesic polygon $P = \Sigma_{i=0}^{n} T(a_i, a_{i+1})$ on a convex surface C which is a Jordan arc. At all interior points of P sides are locally defined; by continuity a right and left side of P can be consistently defined along $P - a_0 - a_{n+1}$. Then $T(a_{i-1}, a_i)$ and $T(a_i, a_{i+1})$, $i = 1, \ldots, n$, form at a_i two sectors to the right and to the left of P. We denote the corresponding angles by α_i and β_i. Consequently,

$$\alpha_i + \beta_i = \text{complete angle at } a_i = 2\pi - \nu(a_i).$$

The numbers

$$s_r(P) = \sum_{i=1}^{n} (\pi - \alpha_i), \quad s_l(P) = \sum_{i=1}^{n} (\pi - \beta_i)$$

correspond to the integral over the geodesic curvature of P towards the right and towards the left of P. If $\nu(a_i) = 0$ for all i, then $s_r(P) = -s_l(P)$ so that on smooth surfaces only one of the two numbers is significant. Alexandrov introduced a brief Russian term (povorot) for this concept which is rendered by "Schwenkung" in German. There does not seem to exist a similarly expressive English term, but we will use "*swerve*" which is brief and has not been previously used as a mathematical term. Thus we call $s_r(P)$ the *right*, and $s_l(P)$ the *left swerve* of P.

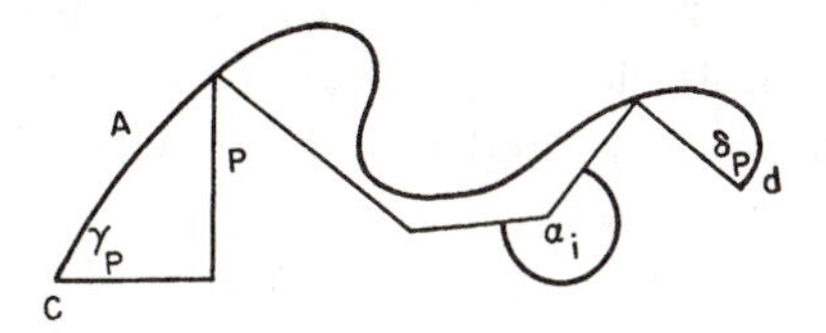

Figure 8

Consider a Jordan arc A which has definite directions at its end points c, d. Take a geodesic polygon P from c to d which is a Jordan arc and lies on the right of, or on, A. This means that any of the regions into which $A \cup P$ decomposes a neighborhood of A (we will let P tend to A) lies to the right of A. As right side of P we denote the side opposite to these regions.

Denote by γ_P and δ_P the angles between P and A at c and

d. Then a comparatively simple application of (14.3), A., pp. 353, 354, yields that

(15.3) $\lim (\gamma_P + \delta_P + s_r(P))$ *exists when* P *approaches* A *from the right,*

i.e., the limit exists and is the same for all sequences P_i which tend to A from the right. This limit is called the *right swerve* $s_r(A)$ of A. The *left swerve* $s_l(A)$ is defined similarly. At first sight it is quite surprising that the existence of definite directions at the end points of A suffices for the existence of $s_r(A)$.

The swerve is evidently additive in the following sense: let the interior point f of A decompose A into A_1 and A_2 and assume that A_1 and A_2 have at f definite directions. Then A_1 and A_2 form an angle α_f to the right of A and

$$s_r(A) = s_r(A_1) + s_r(A_2) + (\pi - \alpha_f).$$

Let A' be the Jordan arc A minus its end points. Clearly, our considerations suggest the relation

$$s_r(A) + s_l(A) = \nu(A'). \tag{15.4}$$

This is indeed correct and a rather simple consequence of the Gauss-Bonnet theorem (15.5).

If A is a closed Jordan curve, then we can approximate A from the right by a simple closed geodesic polygon P. The right side of P is defined as above and it is clear what the right swerve $s_r(P)$ of P means. Then

$s_r(A) = \lim s_r(P)$ *exists, when* P *approaches* A *from the right.*

We call $s_r(A)$ the right swerve of A. This definition and that for Jordan arcs are consistent: if the points c and d of A divide A into Jordan arcs A_1, A_2 which have definite directions at c and d and form the angles α_c and α_d to the right of A, then $s_r(A) = s_r(A_1) + s_r(A_2) + (\pi - \alpha_c) + (\pi - \alpha_d)$.

It is clear that these results in conjunction with (14.3) and the complete additivity of ν yield:

(15.5) GAUSS-BONNET THEOREM. *Denote by* B *the* (by hypothesis non-empty) *interior of a compact set on a convex surface which is bounded by the disjoint closed Jordan curves* $A_1, \ldots, A_k$. *If* $s(A_i)$ *is the swerve of* A_i *towards* B *and* $\chi(B)$ *is the characteristic of* B *then*

$$\nu(B) = 2\pi\chi(B) - \sum_{i=1}^{k} s(A_i).$$

Let A be an arc on the boundary of a *convex* domain $\bar{B}$ with interior B. We want to show that A *has*, as in the plane, *definite directions at its end points* c *and* d. A suitable segment $T(c, x)$ connecting c to a point x of A close to c lies in $\bar{B}$. Then either $T(c, x) \subset A$ or $T(c, x) - c - x$ lies in B. For if $T(c, x)$ contains a point z of B and z' is close to z inside one of the regions in B bounded by A and $T(c, x)$, then $T(c, z')$ and $T(x, z')$ lie by (14.1) in B and have, because of the inclusion property, only c or x in common with $T(c, x)$.

If $T(c, x) \subset A$ then A has evidently a definite direction at c. In the second case, take a point x' on A between x and c and a segment $T(c, x')$ in B and hence in the subdomain of B bounded by $T(c, x)$ and the arc of A from c to x. Again either $T(c, x') \subset A$ and A has obviously a definite direction at c, or we continue. We see that suitable segments $T(c, x)$ with $x \to c$ on A turn monotonically, if at all, and the remark after the proof of (15.1) entails that A has a definite direction at c.

Consider now a polygon $P = \sum_{i=1}^{n} T(a_i, a_{i+1})$, $a_0 = c$, $a_{n+1} = d$ inscribed in A (of course with a_i between a_{i-1} and a_{i+1} on A). The angle α_i of P at a_i towards B is at most π because B is convex, hence the corrsponding swerve of P is non-negative. If P traverses a sequence of polygons with $\max_i a_i a_{i+1} \to 0$ then P tends to A from the side of B and the angle between A and $T(c, a_1)$ or $T(a_n, d)$ tend by the above discussion to 0. Thus we find from the definition (15.3):

(15.6) *The swerve of an arc on the boundary of a convex domain towards the domain is non-negative.*

This and the Gauss-Bonnet theorem entail further:

(15.7) *A convex domain on a convex surface is either a closed convex surface, or homeomorphic to a closed circular disk, or it is isometric to a zone between parallel circles on a circular cylinder.*

A convex domain $\bar{B}$ is, according to our definition, compact and bounded by a finite number of disjoint closed Jordan curves $A_1, \ldots, A_k$. Because $\bar{B}$ is a compact subset of a convex surface it is also a subset of a closed convex surface C. The Gauss-Bonnet theorem applied to the interior B of $\bar{B}$ and (15.6) yield

$$\chi(B) \geqq 0 \textit{ with equality only for } \nu(B) = 0 \textit{ and } s(A_i) = 0.$$

Because C is, topologically, a sphere

$$\chi(B) = 2 - k,$$

hence $k \leqq 2$. The cases $k = 0, 1$ correspond, evidently, to the first two cases in the assertion.

If $k = 2$ then $\nu(B) = 0$ hence the excess of any triangle in B vanishes. Let $a_1a_2a_3$ be such a triangle. With the notations of (13.13) we have $\alpha'_{i+1} = \alpha_{i+1}$ because $\Sigma\alpha_i = \pi = \Sigma\alpha'_i$. If x is

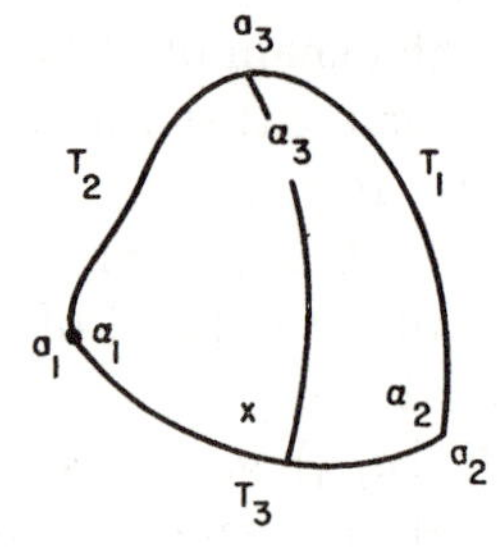

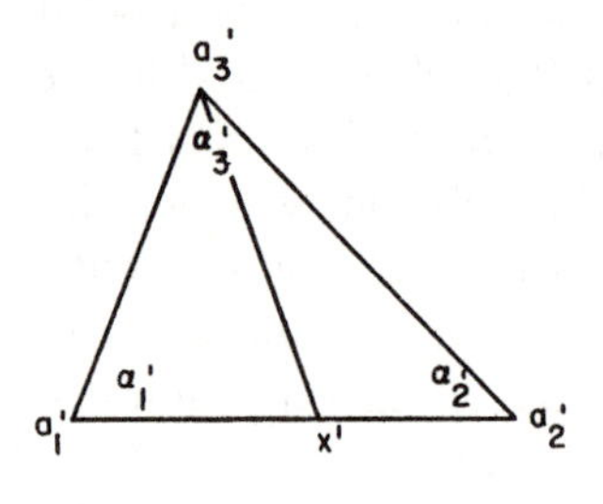

Figure 9

an interior point of T_3 and x' is the point on $E(a'_1, a'_2)$ with $|a'_1 - x'| = a_1x$ then the convexity condition gives $a_1(a_3, a_2) \leqq a_1(a_3, x)$; but (13.9) shows that inequality would imply $\alpha_1 > \alpha'_1$, whence $a_3x = |a'_3 - x'|$.

Repeated application of this fact yields very easily (A., p. 218 may be consulted for the details) that the triangle $a_1a_2a_3$ (boundary and interior), if convex, can be mapped isometrically on the euclidean triangle $a_1'a'_2a'_3$. We know from Section 14 that a given point p of B lies inside a convex polygonal domain D homeomorphic to a disk and contained in B. By diagonals D can be divided into convex triangles. Whether p lies inside one of these triangles or on the boundary of two, it is clear that p has a neighborhood isometric to an open set of E^2. Therefore B is locally isometric to E^2.

We also know from $\chi(B) = 2 - k = 0$ that $s(A_1) = s(A_2) = 0$. On the other hand the swerve of any subarc of A_1 or A_2 toward B is non-negative by (15.6). Consequently the swerve vanishes for every subarc of A_1 or A_2. When B is developed on E^2 such an arc must therefore become a euclidean segment. This implies the assertion for the third case in (15.7).

The geodesics on a convex surface lack many desirable closure properties: A given line element may not lie on any geodesic. If the biconvex lense Λ of Section 12 is approximated by analytic surfaces of revolution then the edge will be the limit of geodesics, but it is itself no geodesic. Also, *there need not be any closed geodesics on a closed convex surface.*

For example, consider a closed convex polyhedron P with the proper vertices $q_1, \ldots, q_n$ so that $0 < \nu(q_i) < 2\pi$. Then $\Sigma \nu(q_i) = 4\pi$. A closed geodesic does not pass through a q_i and decomposes P by the Gauss-Bonnet theorem into two parts each of which has curvature 2π. Therefore the vertices q_i must decompose into two sets for each of which the sum of the curvatures $\nu(q_i)$ equals 2π. Such a decomposition will, in general, not exist, so that there are no closed geodesics on general convex surfaces.

The question arises whether this situation can be remedied by extending the geodesics on arbitrary convex surfaces to a wider class of curves which on smooth surfaces coincides with the set of geodesics. This is possible by using the concept of swerve. A geodesic has the property that for each Jordan arc on it the right and the left swerves vanish. This property is for smooth curves on smooth surfaces sufficient for a curve to be a geodesic, but not on general surfaces. The smoothed lense Λ' of Section 12 contains a curve corresponding to the edge of the lense Λ where every subarc has vanishing right and left swerves. On the lense Λ every subarc of the edge has positive right and left swerves.

Thus we are led to define a *quasigeodesic arc as a Jordan arc which has definite directions at each point and every subarc of which has non-negative right and left swerves.* Because of (15.4) *it suffices to require that the right and left swerves of any subarc have the same sign,* i.e., be both non-negative or both non-positive. A quasigeodesic is a curve which is locally a quasigeodesic arc.

On a convex polyhedron the union $E(a, p_i) \cup E(p_i, b)$ of two segments where p_i is a vertex is a quasigeodesic if the angles between $E(a, p_i)$ and $E(p_i, b)$ at p_i measured on P are both $\leqq \pi$. This shows that the inclusion property is not valid for quasigeodesics. A segment in the sense of the intrinsic metric on a convex surface connecting two conical points with complete angles $\leqq \pi$ (or curvature $\geqq \pi$) forms, traversed back and forth, a degenerate closed quasigeodesic. An edge of a regular tetrahedron furnishes an example.

The quasigeodesics have the desired closure properties:

(15.8) *If the complete convex surfaces C_i, $i = 1, 2, \ldots$, tend to the convex surface C and the quasigeodesic Q_i on C_i tends to the curve Q on C, then Q is a quasigeodesic on C. Moreover, the length of Q_i tends to the length of Q* (see A., p. 374).

In this theorem we assumed tacitly that quasigeodesics are rectifiable. Actually much stronger statements can be made. Pogorelov [2] shows that *Liberman's proof of* (12.1) *extends to quasigeodesics*: if a cylinder Z is constructed through the points of a quasigeodesic arc A as in the beginning of Section 12, see also Figure 7, then after developing Z on a plane the arc A becomes a convex plane curve.

(15.9) *Every line element of a convex surface lies on a quasigeodesic.* (See A., pp. 375, 376).

However, a given line element may now lie on two not partially coinciding quasigeodesics.

Finally, Pogorelov [2] deduces by approximation from the well-known result of Lyusternik-Schnirelman [1] that *any closed convex surface possesses three distinct closed quasigeodesics which are either Jordan curves or quasigeodesics* arcs traversed back and forth.

He also proves that on a surface with bounded specific curvature (the definition is found in Section 4) a quasigeodesic is a geodesic. In that case he therefore obtains three simple closed geodesics.

Much of the theory developed here extends to a wide class of two-dimensional manifolds with an intrinsic metric, see the Conclusion.

CHAPTER IV

Realizations of Intrinsic Metrics

16. The rigidity of convex polyhedra

The present chapter deals with the existence and uniqueness of convex surfaces with given intrinsic metrics.

Let M be a two-dimensional topological manifold with an intrinsic metric $\delta(p', q')$. We say that a convex surface C in E^3 with the intrinsic distance pq *realizes* M, if there is a mapping $p' \rightarrow p$ of M on C such that $\delta(p', q') = pq$ for any two pairs of corresponding points.

If C^0 is congruent to C, that is, if a mapping $p \rightarrow p^0$ of C on C^0 exists which preserves the extrinsic distance: $|p - q| = |p^0 - q^0|$, then C^0 originates from C by a motion of E^3, provided we include reflections among the motions. The intrinsic distances are, of course, also preserved: $pq = p^0q^0$, so that $\delta(p', q') = p^0q^0$, and C^0 is a realization of M. Congruent realizations of M are *not considered as distinct.*

Theorems that certain M have, in this sense, only one realization within a certain class K of convex surfaces, are among the most important accomplishments of the theory, but all languages lack, disturbingly, an adequate term to describe this situation. We will borrow a word from biology and say that M is *monotypic* in K (with corresponding noun monotypy). Explicitly: *M is monotypic in K if at least one realization of M in K exists and any two such realizations are congruent.* We will frequently omit mentioning K when M is monotypic among *all* convex surfaces; notice that two realizations of M are always either both closed, or both open, or both non-complete convex surfaces.

In particular, *M may already be a convex surface in K.* Then the first assumption is trivially satisfied and saying that M is monotypic in K means that a convex surface in K and intrinsically isometric to M is congruent to M. For example, a sphere is monotypic, see end of Section 8, and a convex cap is monotypic among all convex caps, see Section 21.

Our insistence on the convexity of the realization is essential: the surface Z_t, $0 \leqq t \leqq 1$, obtained from the sphere $x_1^2 + x_2^2 + x_3^2 = 1$ by reflecting the cap $x_3 < -1 + t/2$ in the plane $x_3 = -1 + t/2$ is intrinsically isometric, but not congruent, to Z. Thus no statements on monotypy are possible if we leave the realm of convex surfaces. The surfaces of class C^m, $m \geqq 2$, with positive Gauss curvature occurring in the classical theorems, are automatically convex.

A convex surface C is *deformable* in a certain class K of convex surfaces if a family C_t, $0 \leqq t \leqq 1$ of surfaces in K exists with the following properties: 1) For each t there is a mapping $p \rightarrow p_t$ of C on C_t which preserves the intrinsic distances $pq = p_t q_t$; 2) $p_0 = p$; 3) p_t depends continuously on t; 4) $p \rightarrow p_1$ is not a congruence, i.e., $|p - q| \neq |p_1 - q_1|$ for certain pairs p, q on C. We also describe C_t as a (proper) deformation of C into C_1 in K.

It follows from 1) and 3) that p_t is continuous on $C \times [0, 1]$. The above example Z_t shows that the restriction to convex surfaces is essential also in this case.

A monotypic convex surface is not deformable; whether the converse holds does not seem to be known in complete generality. But we will see that it holds for large classes of convex surfaces.

Originally, a non-deformable surface was called rigid (in the class considered). It is now customary to use the word *rigid for what was formerly called infinitesimally rigid.* Since we will not use this term, except for polyhedra among polyhedra, we only describe it loosely for the general case: C is rigid if any variation C_t of $C = C_0$ which leaves the lengths of all curves on C stationary for $t = 0$ also leaves the distances $|p_t - q_t|$ stationary for $t = 0$. [1]

The condition $d|p_t - q_t|/dt_{t=0} = 0$ implies that the velocity vectors $\mathfrak{v}(p) = dp/dt_{t=0}$ coincide with those of a *rigid motion*. Therefore,

[1] For exact formulations in the classical case under minimal conditions see Minagawa and Radó [1].

if $\mathfrak{r}(p)$ is the radius vector from z to p, then two fixed vectors $\mathfrak{a}$, $\mathfrak{b}$ exist such that (on C)

$$\mathfrak{v}(p) = \mathfrak{a} + \mathfrak{b} \times \mathfrak{r}(p). \tag{16.1}$$

For this reason we call a vector field (a vector field is by definition continuous) $\mathfrak{v}(p)$ defined on all points of a set M, *trivial* (on M) if it can be represented in the form (16.1).

We now turn to polyhedra, and since we no longer consider only closed polyhedra, we define: *A (complete) convex polyhedron P* in E^3 is a complete convex surface which is contained in a finite number of planes.[2] It evidently suffices to consider planes whose intersections with P have dimension 2. These intersections are the *(natural) faces* of P. There is a finite number of *(natural) edges* bounding these faces, which are either segments, rays, or straight lines, and a finite number of *natural vertices or corners.* The term "natural" will be used in contradistinction to certain other faces and edges which we will have to admit in our considerations.

There are three types of polyhedra: closed, open, and cylindrical. The latter will not play any role. Owing to the fact that a doubly-covered plane convex domain (with a boundary) is considered as a convex surface, we also consider doubly covered plane convex polygonal regions with a non-empty boundary as convex polyhedra. Such polyhedra are called *degenerate.*

A non-complete convex polyhedron is a connected open subset of a complete polyhedron whose relative boundary lies on a finite number of straight lines.

As is well known, the first monotypy theorem is due to Cauchy who proved in 1813, that two closed convex polyhedra are congruent if there is a one-to-one incidence preserving correspondence between their natural faces, edges, and vertices respectively and corresponding faces are congruent.[3] Since such polyhedra are

[2] Notice that for unbounded polyhedra this definition deviates from that customary in topology. Complete convex surface is meant as on page 3.

[3] Proofs of Cauchy's theorem are available in many places, for example, in Steinitz and Rademacher [1], and A. [8].

intrinsically isometric, this theorem now appears as a very special case of Pogorelov's theorem that *any closed convex surface is monotypic.* However, Porogelov uses Cauchy's theorem in his proof (see Section 21), and Cauchy's method contributed an essential idea to the first monotypy theorem on curved convex surfaces by Cohn-Vossen; moreover, it yields with hardly any change the rigidity of convex polyhedra,[4] which in view of our preceding remark may be formulated as follows:

(16.2) *If a vector field* $\mathfrak{v}(p)$ *is defined on a closed convex polyhedron P and is trivial on each natural face of P then it is trivial on all of P.*

Actually this formulation of rigidity implies a sort of converse of the previous remark. But, under the hypothesis of (16.2) and with the notations of (16.1), $\mathfrak{r}(p) + t\mathfrak{v}(p)$ represents a variation P_t of P which, at least for small t, is a convex polyhedron. A face of P_t corresponds to a face of P and each face is moved rigidly at the moment $t = 0$.

We are here interested in generalizations of (16.2) in two directions: the first concerns *non-closed polyhedra* and is considered for its own interest, the second concerns a *wider concept of face* and is necessary for an important application.

We subdivide a natural face into "*conditional*" faces by diagonals which do not intersect in the interior of the face. For an unbounded face we allow as diagonals also a finite number of rays emanating from vertices (and not intersecting in the interior) of the face.

(16.2) *ceases to be valid for degenerate polyhedra* (except triangles), if we only assume that $\mathfrak{v}(p)$ is trivial on the conditional faces. For example, if for a doubly covered square *abcd* with side 1 we consider the diagonals $E(a, c)$ and $E(b, d)$ as drawn on different faces of the square and lift the vertex c a distance t vertically out of the plane of the square, then $|a - b|$, $|b - d|$, $|a - d|$ do not change and $|b - c| = |c - d| = (1 + t^2)^{1/2}$, $|a - c| = (2 + t^2)^{1/2}$ so that the derivatives of the lengths of these edges vanish for $t = 0$.

[4] This fact was first proved explicitly by Dehn [1], who overlooked the applicability of Cauchy's argument.

However, (16.2) *remains valid for non-degenerate polyhedra.* The generalization is considerable: we permit folding of the natural faces along certain diagonals and this folding may proceed "inward" or "outward," hence would in the first case destroy convexity. With the restriction to "outward" folding, i.e., to convex surfaces, Alexandrov proves this theorem in A., pp. 239–253. In [8], Chapter X[5] he proves the general theorem and goes one step farther by allowing new vertices to appear on the natural edges. (It is obvious that new vertices in the interior of the natural faces would destroy rigidity.) He also proves (16.2) for open polyhedra with curvature 2π and for certain polyhedra with a boundary. His methods constitute a refinement of Cauchy's. Without allowing new vertices we will follow here a method developed by Pogorelov [5].

A convex polyhedral angle A is a non-degenerate open convex polyhedron with exactly one vertex a and hence at least three natural faces. As stated before, we allow subdivision of these faces into a finite number of conditional faces by rays emanating from a. The principal tool of Pogorelov is the following lemma:

(16.3) *Let* $\mathfrak{v}(p)$ *be a vector field on* A *with* $\mathfrak{v}(a) = 0$ *which is trivial on each conditional face of* A*; let* R *be a ray from* a *in the interior of* A *(except for* a*). If the component of* $\mathfrak{v}(p)$ *in the direction of* R *is positive (negative) on one natural edge (except for* a*) then it is negative (positive) on at least one natural edge.*

We place the origin at a and take R as positive x_3-axis. A vector field $\mathfrak{v}(p)$ which is trivial on a face of A has, because of $\mathfrak{v}(a) = 0$, the form $\mathfrak{b} \times \mathfrak{x}(p)$, hence $v_3(p)$ either vanishes on a ray with origin a or has for $p \neq a$ the same sign. Our proof is indirect: we assume the existence of a vector field $\mathfrak{v}(p)$ on A with $\mathfrak{v}(a) = 0$, trivial on each conditional face, for which $v_3(p) \geqq 0$ on each natural edge and $v_3(p) > 0$ for $p \neq a$ on at least one.

We denote the natural edges of A in cyclical order by

[5] This book presupposes very little knowledge and is written in great detail with much emphasis on motivation. It provides a wealth of information as well as real fun to anyone who reads Russian and has more than the minimum amount of knowledge required.

$E_1, \ldots, E_n$, $n \geqq 3$, $E_{n+k} = E_k$ and by F_i the natural face with $E_i \cup E_{i+1}$ as boundary. Then $\mathfrak{v}(p)$ has on F_i the form $\mathfrak{b}^i \times \mathfrak{x}(p)$. Without changing this vector field on F_i we may replace $\mathfrak{b}^i$ by a vector $\mathfrak{b}$ in the intersection of the planes spanned by $\mathfrak{b}^i$, E_i and $\mathfrak{b}^{i+1}$, E_{i+1} respectively. Therefore we can define linearly a new vector field $\mathfrak{v}'(p)$ on F_i which coincides with $\mathfrak{v}(p)$ on E_i and on E_{i+1} moreover $v_3(p) \geqq 0$ on F_i. Then $\mathfrak{v}'(p)$ is trivial on F_i. In this way we obtain *a vector field* $\mathfrak{v}'(p)$ *on all of* A *which is trivial on all natural faces and satisfies* $v_3'(p) \geqq 0$ *and* $v'_3(p) > 0$ *for* $p \neq a$ *on at least one* E_i.

Next we show that we may even assume $v'_3(p) > 0$ *for* $p \in A - a$. If $v'_3(p) > 0$ on each $E_i - a$, then the linearity of $\mathfrak{v}'(p)$ on F_i shows that $v'_3(p) > 0$ on $F_i - a$.

Assume $v'_3(p) > 0$ on $E_2 - a$ but $v_3'(p) = 0$ on E_3. We then vary the part of A formed by F_1, F_2, F_3 such that E_1 and E_4 remain fixed and the angles of F_1, F_2, F_3 at a do not change. We denote by $\mathfrak{u}$ the velocity vector of this variation. Since along E_3 this is the field of a rotation about E_4, we will have $v'_3 + \epsilon u_3 > 0$ on $E_3 - a$ for a suitable choice of sign ϵ, and if $|\epsilon|$ is small enough $v'_3 + \epsilon u_3 > 0$ on $E_2 - a$. Thus we obtain step by step a vector field on A which we denote again by $\mathfrak{v}(p)$ with $v_3(p) > 0$ for $p \neq a$, and trivial on each F_i.

The vector field $\mathfrak{v}(p)$ *is not trivial on all of* A. For if it were, it would belong to a rotation about an axis passing through a because $\mathfrak{v}(a) = 0$. The plane through this axis and R would intersect A in two rays along which $v_3(p) = 0$.

As in the remark after (16.2) we consider the variation $\mathfrak{x}(p) + t\mathfrak{v}(p)$ of A and show that the *angles* $\alpha_i = \sphericalangle (F_{i-1}, F_i)$ *cannot all increase or all decrease*. In the proof we may assume $\mathfrak{v}(p) = 0$ on F_1 and denote by $\mathfrak{w}^i$ the trivial vector field on E^3 which coincides with $\mathfrak{v}(p)$ on F_i, hence $\mathfrak{w}^1 \equiv 0$. Since $\mathfrak{v}(a) = 0$, the field $\mathfrak{w}^i$ belongs to a rotation about a line through a and $\mathfrak{w}^i - \mathfrak{w}^{i-1}$ belongs to a rotation about E_i. Since

$$\mathfrak{w}^n = (\mathfrak{w}^n - \mathfrak{w}^{n-1}) + (\mathfrak{w}^{n-1} - \mathfrak{w}^{n-2}) + \ldots + (\mathfrak{w}^2 - \mathfrak{w}^1)$$

we may regard $\mathfrak{v}(p)$ for $p \in E_1 - a$ as composed of rotations about $E_n, E_{n-1}, \ldots, E_2$. If all α_i increased or all α_i decreased then all the vectors $\mathfrak{w}^i(p) - \mathfrak{w}^{i-1}(p)$ would point into one half space, hence their sum can vanish only if each vanishes. This would have to be the case because E_1 stays fixed. But then $\mathfrak{v}(p)$ would be trivial on A, which is contrary to our previous result.

It follows that either there is an angle $\angle (E_i, E_{i+2})$ which is stationary or there are consecutive angles $\angle (E_i, E_{i+2})$ and $\angle (E_{i+1}, E_{i+3})$ one of which increases and the other decreases.

If $\angle (E_i, E_{i+k})$, $k \geqq 2$, is stationary then the plane through E_i and E_{i+k} divides A into two polyhedral angles A', A''. On the common face of A' and A'' bounded by E_i and E_{i+k} we can define $\mathfrak{v}'$ linearly as above so that it coincides with $\mathfrak{v}$ on E_i and E_{i+k}.

Assume now that no $\angle (E_i, E_{i+k})$, $k \geqq 2$, is stationary and that $\angle (E_i, E_{i+2})$ increases whereas $\angle (E_{i+1}, E_{i+3})$ decreases. If $\angle (E_i, E_{i+3})$ decreases then there is a ray E from a in F_{i+2} such that $\angle (E_i, E)$ is stationary. If $\angle (E_i, E_{i+3})$ increases then $\angle (E_{i+3}, E)$ will be stationary for a suitable ray E on F_i. We divide A into two angles A', A'' in the first case by the plane through E_i and E, in the second by the plane through E_{i+3} and E, and proceed as above.

In all three cases we obtain angles A', A'' with fewer faces than A, and on A' and on A'' a vector field $\mathfrak{v}(p)$ with $v_3(p) > 0$ for $p \neq a$ and trivial on the natural faces. Eventually we obtain a polyhedral angle with three faces which is obviously rigid, but we know that $\mathfrak{v}(p)$ cannot be trivial on the whole angle.

We use this lemma to establish the rigidity of caps. According to our previous definition a convex cap which is a polyhedron cannot contain faces perpendicular to the plane containing the boundary of the cap. For an important application we *enlarge* the concept of cap to include this case, formally: *A convex polyhedral cap P is a convex polyhedron homeomorphic to an open disk whose boundary B is a closed plane convex polygon such that a perpendicular to the plane Q of B through a point of P cuts Q inside or on B.*

As conditional faces we also admit faces obtained from a (possibly already conditional) face of P perpendicular to Q by subdivision with perpendiculars to Q through vertices of P. A cap is rigid when its boundary is constrained to stay in its plane. Precisely:

(16.4) THEOREM. *A vector field* $\mathfrak{v}(p)$ *defined on the closure* $\bar{P}$ *of a convex polyhedral cap* P *which is trivial on each conditional face of* P *and along the boundary* B *of* P *lies in the plane* Q *of* B, *is trivial on all of* P.

We choose Q as (x_1, x_2)-plane and assume that $\bar{P}$ lies in $x_3 \geqq 0$. Denote by P' the part of P formed by those faces which are not perpendicular to Q.

We first consider the case where $v_3(p) \equiv 0$ *on* P. Let F' and F'' be two faces of P' with a common edge E and $\mathfrak{w}'$, $\mathfrak{w}''$ the trivial vector fields in E^3 coinciding with v on F', F'' respectively. Then, as above, $\mathfrak{w}' - \mathfrak{w}'' = 0$ on E and $\mathfrak{w}' - \mathfrak{w}''$ belongs to a rotation about E. Since E is not perpendicular to Q, it follows that $w'_3(p) - w''_3(p) = 0$ on $F' \cup F''$ implies $\mathfrak{w}'(p) - \mathfrak{w}''(p) = 0$ on $F' \cup F''$. Thus $\mathfrak{v}(p)$ is trivial on $F' \cup F''$ and on all of P'.

Denote by $\mathfrak{w}$ the trivial vector field in E^3 coinciding with $\mathfrak{v}$ on P'. Clearly $\mathfrak{v} = \mathfrak{w}$ also on those faces perpendicular to Q whose boundaries have no proper segment in Q, because all their vertices lie in P'. (Such faces may be present through subdivision of natural faces by diagonals.)

We want to show $\mathfrak{v} = \mathfrak{w}$ on all of P. It suffices to establish $\mathfrak{v}(b) = \mathfrak{w}(b)$ for all vertices of B not in P' if we include among them the end points of conditional edges perpendicular to Q. Each b lies on an edge E of P perpendicular to Q whose other end point c lies in P'. Let F', F'' be the two faces of P with E as common edge and denote by E' and E'' the other edges of F' and F'' with end point c.

First take the case where E is a natural edge of P. Then $\mathfrak{v} - \mathfrak{w}$ is on F' and F'' a field of rotation about E' or E''. Hence $\mathfrak{v}(b) - \mathfrak{w}(b)$, if different from 0, would be perpendicular to the three non-coplanar edges E, E', E''.

Assume next that E is a conditional edge but that the other edge of F' perpendicular to Q is a natural edge. Then we know that $\mathfrak{v} = \mathfrak{w}$ on all of F', in particular on E and also on E'', hence

$\mathfrak{v} = \mathfrak{w}$ on all of F''. The same argument yields then that $\mathfrak{v} = \mathfrak{w}$ on the other conditional face perpendicular to Q adjacent to F'' (if it exists) and thus, step by step, $\mathfrak{v} = \mathfrak{w}$ on all of P.

We now consider any vector field $\mathfrak{v}$ *satisfying the hypothesis of* (16.4). We show first that $v_3(p) = 0$ on the boundary of P'. It suffices to prove $v_3(p) = 0$ for every vertex of P on the boundary of P'. If q lies also on B then $v_3(q) = 0$ by hypotheses. If $q \in B$ let f be the foot of q on Q. If $E(q, f)$ lies in one face of P then $v_3(f) = 0$ entails $v_3(q) = 0$ because $\mathfrak{v}$ corresponds to a rigid motion of this face.

In general $E(q, f)$ will not lie in one conditional face only, but will cross conditional edges. If it crosses only one at f_1, say, then v_3 vanishes at the end points of this edge, as we just saw, hence $v_3(f_1) = 0$ and we conclude as before that $v_3(q) = 0$. In the same way, the case where $E(q, f)$ crosses two conditional edges is reduced to the case where it crosses one, and so forth.

If the theorem were false, then the first part of this proof shows that $v_3(p)$ cannot vanish on all of P, hence $|v_3(p)|$ would attain a maximum $\alpha > 0$. If this maximum is attained in the interior of a face or edge, then $|v_3(p)| = \alpha$ on the whole face or edge, hence there is a vertex q of P with $|v_3(q)| = \alpha$. We know that q must be an interior point of P'. We connect q to a vertex on the boundary of P' by a path consisting of natural edges. Denote by q' the first vertex following q on this path for which $|v_3| < \alpha$ and by a the vertex preceding q' so that $|v_3(a)| = \alpha$ and $|v_3(q')| < \alpha$.

Let A be the polyhedral angle with vertex a whose faces contain the faces of P contiguous to a, and continue the vector field $\mathfrak{v}(p)$ over all of A in the obvious way, namely so that it becomes trivial on each face of A. Then $|v_3(p)|$ attains also on A its maximum at a. If R is the ray parallel to the negative x_3-axis with origin a, then the vector field $\mathfrak{w}(p) = \mathfrak{v}(p) - \mathfrak{v}(a)$ satisfies the hypotheses of (16.3) because $w_3(q') \neq 0$. Consequently, $w_3(p)$ changes sign on A, but this would mean that $|v_3(p)|$ does not attain a maximum at a.

That *open convex polyhedra are in general not rigid* is obvious: no convex polyhedral angle with more than three faces is rigid.

However:

(16.5) THEOREM. *A non-degenerate open convex polyhedron P with total curvature $\nu(P) = 2\pi$* is *rigid. Or, a vector field trivial on all conditional faces of P is trivial on all of P.*

Since $\nu(P) = 2\pi$ the unbounded edges of P are parallel rays pointing into the same half space. Choose a plane Q normal to these rays such that it intersects all these rays. We remember that the natural faces of P perpendicular to Q may be divided into conditional faces by diagonals and by rays from vertices of P perpendicular to Q, but these diagonals and rays must not intersect in the interior of a face.

Q divides P into a cap C and a half cylinder. Denote the unbounded faces of P in cyclical order by $F_1, \ldots, F_n, F_{n+k} = F_k$, and let E_i be the unbounded edge or ray common to F_{i-1} and F_i. If $\mathfrak{w}^i$ is the trivial vector field in E^3 coinciding with $\mathfrak{v}$ on F_i, then $\mathfrak{w}^i - \mathfrak{w}^{i-1}$ vanishes on E_i, in particular its component in the direction of E_i, i.e., normal to Q. Therefore this component vanishes on $F_{i-1} \cup F_i$. Now

$$\mathfrak{w}^i - \mathfrak{w}^1 = (\mathfrak{w}^i - \mathfrak{w}^{i-1}) + \ldots + (\mathfrak{w}^2 - \mathfrak{w}^1)$$

shows that the component of $\mathfrak{w}^i - \mathfrak{w}^1$ normal to Q vanishes. Therefore the component of $\mathfrak{v} - \mathfrak{w}^1$ normal to Q vanishes on all unbounded faces of P. Thus $\mathfrak{v} - \mathfrak{w}^1$ satisfies on the cap C the hypotheses of the last theorem. Therefore $\mathfrak{v} - \mathfrak{w}^1$ is trivial on C, hence on all of P. Since $\mathfrak{w}^1$ is trivial, $\mathfrak{v}$ is too.

Various classes of open polyhedra with curvature less than 2π are known within which a polyhedron is rigid, but we will not discuss them here and merely refer the reader to A. [8], pp. 400–403 and to Pogorelov [5].

Finally we consider closed polyhedra and prove the theorem which we will need in the next section:

(16.6) THEOREM. *A non-degenerate closed convex polyhedron P is rigid. Or, a vector field $\mathfrak{v}(p)$ on P which is trivial on each conditional face of P, is trivial on all of P.*

By a well-known idea of Darboux [1], Livre VIII, Chapitre III we reduce this theorem to the preceding one. We take a supporting plane Q of P which has a vertex z of P and no other point in common with P. We choose Q as (x_1, x_2)-plane and z as origin of a rectangular coordinate system x_1, x_2, x_3. Darboux's transformation

(16.7) $x'_1 = x_1/x_3, \quad x'_2 = x_2/x_3, \quad x'_3 = 1/x_3$

takes P into an open polyhedron P' whose unbounded edges are the transforms of the natural and conditional edges with z as one end point and are normal to Q. Therefore P' has curvature 2π. The fundamental property of (16.7) is that the vector field $\mathfrak{v}$ is trivial on the set M, if and only if, the vector field $\mathfrak{v}'$ defined by

$$v'_1 = v_1/x_3, \; v'_2 = v_2/x_3, \; v'_3 = -(x_1v_1 + x_2v_2 + x_3v_3)/x_3$$

is trivial on the image of M under (16.7). This is easily verified, but may also be found in Darboux (*loc.cit.*). Therefore the image $\mathfrak{v}'$ on P' of the field $\mathfrak{v}$ on P satisfies the hypotheses of the last theorem and hence is trivial on all of P'. It follows that $\mathfrak{v}$ is trivial on all of P.

17. The realization of polyhedral metrics

We now turn to realization problems. The problem for closed convex surfaces will be solved for polyhedra and then by approximation for general convex surfaces.

First we must therefore recognize the *intrinsic characteristic of a metric on a polyhedron.* Every point of a (two-dimensional) polyhedron, except for the vertices, has a neighborhood which is isometric to an open set (of circular disk) in E^2. This applies even to interior points of edges, because the union of two distinct half planes with the same boundary can be unfolded on a plane. The neighborhood of a vertex is isometric to a neighborhood of the apex of a suitable cone. If the polyhedron is convex then the complete angle of this cone is less than 2π.

To define *conical neighborhood* with an arbitrary angle exactly, take a closed rectifiable Jordan curve A on the unit sphere Z in E^3. The rays $R(z, x)$ form the center z of Z to a variable point x on A form a *cone* B, *whose intrinsic metric depends exclusively on the length* α *of* A. We call α the *complete angle of* B (at z) and $2\pi - \alpha$ the *curvature of* B. If B is cut open along a generator and unfolded in the (x_1, x_2)-plane so that the apex falls into (0, 0) the curve A falls on the unitcircle $x_1^2 + x_2^2 = 1$ and α indicates how large a portion of this circle is traversed (it may be any positive number). The metric on B is euclidean if $\alpha = 2\pi$.

A *polyhedral metric* is an intrinsic metric on a two-dimensional manifold M such that every point p of M has a neighborhood which can be mapped isometrically on a neighborhood of the apex of a cone with p going into the apex. The curvature of the cone is the *curvature of M, or of the metric, at p.* It should be noticed that the second condition alone does not imply intrinsicness: min $(1, |x - y|)$ is a metrization of E^2 satisfying the second condition, but which is not intrinsic.

The intrinsic metric of a closed convex polyhedron in E^3, including a doubly covered plane domain bounded by a closed convex polygon, is a polyhedral metric on S^2 with non-positive curvature. It is one of the fundamental results of Alexandrov that every such metric on S^2 is uniquely realizable as a polyhedron, or in our brief terminology: is monotypic among the convex polyhedra, actually among all convex surfaces. Because of its importance we formulate the result at length:

(17.1) THEOREM. *A polyhedral metric with non-positive curvature on the sphere can be realized as one, and up to motions only one, (possibly degenerate) polyhedron.*

The uniqueness among polyhedra is somewhat more general than Cauchy's theorem, but can be proved by Cauchy's method. Olovyanishnikov [1] deduces uniqueness among all convex surfaces from (14.7) (or rather from a special case of (14.7) proved by him; (14.7) was then not known). This result is contained in the monotypy theorem (21.1), whose present proof uses, however, uniqueness in (17.1).

The existence proof for (17.1) is quite long when carried out in all details, it is found in A., Chapter VI. Here we outline the proof (in greater detail than Alexandrov's own sketch, *loc. cit.*).

Since the apex of a cone is the only point of the cone which does not have a euclidean neighborhood, the points with non-vanishing curvature of a polyhedral metric are isolated. We call them *corners.* A polyhedral metric with non-negative curvature on S^2 will in this section, for brevity, be denoted as *"Metric,"* in particular as *c-Metric,* if there are c corners $a_1, \ldots, a_c$. Let $\alpha_i > 0$ be the curvature of a_i.

We know from our discussion in Section 11 that an interior point of a segment cannot be a corner, and hence has a euclidean neighborhood. This yields at once the *inclusion property* (13.10) and (13.11): If (abc) then $T(a, b)$ and $T(b, c)$ are unique. (Segments exist because S^2 is compact.) It is also evident that the *complete angle at each point exists* and equals 2π minus the curvature. A segment is divided by an interior point into two segments forming two angles of measure π. Thus the general theory developed in Sections 13, 14 is applicable although there are, of course, much simpler arguments covering the present case. We merely want to emphasize that we stay within our general theory.

The total curvature equals 4π, hence there are at least three corners, $c \geqq 3$. We begin by constructing *particularly simple triangulations* for a given Metric M on S^2. Draw segments $T_2, \ldots, T_c$ from a_1 to the other corners $a_2, \ldots, a_c$. Since T_i does not pass through any corner, we conclude from (13.11) that $T_i \cap T_j = a_1$ for $i \neq j$. We cut the sphere open along $T_2, \ldots, T_c$ and obtain a polygon Q with $2c - 2 = m$ vertices $b_1, \ldots, b_m$ corresponding to the a_c in the cyclical order $a_1 a_2 a_1 a_3 a_1 \ldots$. Let β_i be the angle inside Q at b_i. Then $\beta_1 + \beta_3 + \ldots + \beta_{2c-3} = 2\pi - \alpha_1$ and $\beta_2 + \beta_4 + \ldots + \beta_{2c-2} = (2\pi - \alpha_2) + (2\pi - \alpha_3) + \ldots + (2\pi - \alpha_c)$, hence $\Sigma\, \alpha_i = 4\pi$ yields $\Sigma\, \beta_i = (m - 2)\pi$. This could also have been concluded from the fact that Q contains no corner in its interior and behaves like a plane polygon.

Therefore $\beta_i < \pi$ for at least three i, and there are two vertices b_p, b_q with $\beta_p < \pi$, $\beta_q < \pi$ separated by two other vertices b_k and b_l. A shortest join C of b_k and b_l in Q cannot pass through a point b_i with $\beta_i < \pi$ because it could then be shortened. For the same reason and because of the inclusion property C cannot contain an interior point of a side of Q without containing the whole side. On the other hand, C separates b_p from b_q in Q, hence must partially lie inside Q. Thus it contains an arc D which connects two vertices b_r and b_s of Q and lies otherwise inside Q. For M the arc D is a geodesic. It decomposes Q into two polygons with fewer verticses, to each of which we can apply the same procedure, until we have obtained *a triangulation of M into geodesic triangles all whose vertices are corners a_i.*

Each of these triangles is isometric to a triangle in E^2. For if abc is one of the triangles and A the side opposite a, then a shortest join C_x of a to a point x on A in abc lies, as we just saw, except for a and x, inside abc, moreover C_x is unique. For, two shortest joins would, by (13.11), bound a simply connected domain in abc with positive curvature ν by the Gauss-Bonnet theorem.

Since C_x is a geodesic on M, every point of $C_x - a$ has a euclidean neighborhood. Using also, that a neighborhood of a is isometric to the neighborhood of the apex of a cone, we find that a given point x on A is interior point of a subarc A' of A such that

$$\bigcup_{x \in A'} C_x$$

is isometric to a triangle in E^2. The Heine-Borel theorem yields the same for all of abc.

If $c = 3$ then connecting the vertices a_1, a_2, a_3 by segments decomposes M into two triangles each isometric to a plane triangle a_1, a_2, a_3 with $|a_i - a_k| = a_i a_k$. Thus

(17.2) *A 3-Metric is realized by a doubly covered plane triangle.*

This is important for us because the existence proof will progress by induction with respect to c: we show that a given c-Metric can be realized provided all $(c-1)$-Metrics, $c - 1 \geqq 3$, can be realized, with the following main steps. Calling a Metric *degenerate* if it is realizable as a degenerate convex polyhedron and realizable if it is realizable as a convex polyhedron in E^3 (notice that degenerate Metrics are realizable by definition) we show:

I. *A c-Metric sufficiently close to a non-degenerate c-Metric is realizable.*

II. *If a sequence of realizable c-Metrics converges to a Metric then the latter is realizable by a limit of a suitable subsequence of polyhedra realizing the Metrics of the sequence.*

III. *If every non-degenerate* $(c-1)$*-Metric,* $c-1 \geqq 3$*, is realizable then a given non-degenerate c-Metric* ϱ_0 *lies in a continuous family* ϱ_t, $0 \leqq t \leqq 1$, *of non-degenerate c-Metrics where* ϱ_1 *is realizable.*

A triangulation of a c-Metric M into geodesic triangles all whose vertices are corners will be called a *(proper) net* N. If N has f faces and e edges then $3f - 3e + 3c = 6$ by Euler's theorem and $3f = 2e$ because all faces are triangles, hence

$$e = 3c - 6, \quad c = (e + 6)/3. \tag{17.3}$$

We say two nets N_1, N_2, not necessarily on the same manifold, have the *same structure* if there is a one-to-one incidence preserving mapping of the faces, edges and vertices of N_1 on the faces, edges, and vertices, respectively, of N_2. We will use the term "corresponding" faces, . . . of N_1 and N_2 somewhat loosely for corresponding elements under such a mapping, although the mapping is, in general, not unique. The context will imply which correspondence is meant.

A closed polyhedron P in E^3 can be triangulated into a net by dividing its natural faces into triangles by diagonals which do not intersect in the interior of the faces. These diagonals and the newly created faces are conditional edges and faces in the terminology of the last section. A net on P obtained in this way will be called *extrinsic*. With respect to its intrinsic metric P will, in general, also have nets which are not extrinsic.

A convex polyhedron P in E^3 is known when its vertices or their $3c$ coordinates are known. Changing the position of the vertices will change the shape of P unless the change happens to correspond to a rigid motion. We eliminate this trivial possibility by selecting 3 vertices a, b, c of P and fixed rectangular coordinates x_1, x_2, x_3 and placing P so that a falls in $z = (0, 0, 0)$, $R(a, b)$ on the positive x_1-axis and c into the half plane $x_3 = 0$, $x_2 > 0$. This eliminates 6 parameters, so that *the shape of* P *depends on* $3c - 6$ *parameters.* It is most important for our argument that this equals the number e of edges of a proper net on P.

Consider a polyhedron P^0 with c vertices. If these vertices are moved continuously but so that the resulting polyhedron P is convex, then the faces of P^0 may be folded along certain diagonals. For P close to P^0 the number of vertices is still c, hence there corresponds to an extrinsic net E_E on P a definite extrinsic net N^0 on P^0. For different P even arbitrarily close to P^0, the nets $N_p{}^0$ may be different, but there is only a finite number of extrinsic nets on P^0.

Take any not necessarily extrinsic net N^0 on P^0. The segment T connecting two vertices a_i, a_j of P^0 is in general, not unique, but if two segments exist, then each of the two domains on P^0 bounded by them, by the Gauss-Bonnet theorem, contains a corner of P^0 in its interior. T is the only segment $T(a_i, a_j)$ in any neighborhood U of T which does not contain other vertices than a_i and a_j. This, and the continuous dependence of the intrinsic metric on P, see (11.2), shows that for P close to P^0, there is exactly one net N of P close to N^0 and of the same structure as N^0.

Denote by r_i and $r_i{}^0$, $i = 1, \ldots, e$, the lengths of corresponding edges in N and N^0. If three vertices abc of P^0 and the corresponding vertices of P are placed as above and $p = (p_1, \ldots, p_e)$, $p^0 = (p_1{}^0, \ldots, p_e{}^0)$ are the $3c-6$ free coordinates of the vertices of P and P^0, then

$$r_i = f_i(p) = f_i(p_1, \ldots, p_e),\ i = 1, \ldots, e,$$

are continuous functions of p in a neighborhood of p^0. Actually they are *continuously differentiable*, A., pp. 239–242, (this is not quite as trivial as it looks because certain edges may be broken). If N_E is an extrinsic net on P which varies continuously with P, then the lengths $s_1, \ldots, s_e$ of its edges are functions of class C^1 of the $r_1, \ldots, r_e$, A., p. 230, so that

$$ds_i = \sum_j \frac{\partial s_i(r^0)}{\partial r_j} dr_j,\ i = 1, \ldots, e,$$

and $dr_j = 0$ entails $ds_i = 0$. But the condition $ds_i = 0$ means that the conditional edges s_i of P are stationary for $r = r^0$. If P^0 is non-degenerate, then our rigidity theorem (16.6) tells us that the variation of P^0 for $r = r^0$ amounts to a rigid motion.

Since we fixed abc this means that $dp_i = 0$. Thus, in the equations

$$dr_i = \sum_j \frac{\partial f_i(p^0)}{\partial p_j} dp_j, \quad i = 1, \ldots, e,$$

if $r_i = 0$, $i = 1, \ldots, e$, then $dp_i = 0$, $i = 1, \ldots, e$. Therefore the *determinant* $|\partial f_i(p^0)/\partial p_j|_{i,j}$ *does not vanish,* and the equations $r_i = f_i(p)$ can be inverted:

$$p_i = g_i(r_1, \ldots, r_e) \text{ in a neighborhood of } r^0.$$

In the argument we excluded the possibility that the extrinsic net N_E on P does not vary continuously, which it definitely may. This contingency is easily taken care of because the number of extrinsic nets on P^0 is finite.

Thus we have accomplished step I of our program:

(17.4) *Let the c-Matric with the net* N^0 *and edge lengths* $r_1{}^0, \ldots, r_e{}^0$ *be realizable and non-degenerate. Then an* $\epsilon > 0$ *exists such that any c-Metric with a net* N *of the same structure as* N^0 *and lengths* $r_i{}^0$ *and* r_i *of corresponding edges satisfying* $|r_i - r_i{}^0| < \epsilon, i = 1, \ldots, e,$ *is also realizable.*

We omit the proof of II. It is simple, but is not an immediate consequence of (11.2), because the net N and the lengths r_i of its edges enter (17.4), so that II must be formulated in these terms too, see A., pp. 255, 256.

The *methods* make step III very interesting. In particular the method of "*cutting and gluing*" (so called by Alexandrov), which enters the first argument, has proved a powerful tool in many other investigations.

(17.5) *Any c-Metric* ϱ_0, $c \geqq 4$, *can be imbedded in a continuous family* ϱ_t, $0 \leqq t \leqq 1$, *of Metrics such that all* ϱ_t *with* $t < 1$ *are c-Metrics and* ϱ_1 *is a* $(c-1)$*-Metric.*

Since (in the Metric ϱ_0) $c \geqq 4$ and $\Sigma\, \alpha_i = 4\pi$, there are two α_i such that $\alpha_i + \alpha_j < 2\pi$, unless $c = 4$ and all $\alpha_i = \pi$.

Assume $\alpha_1 + \alpha_2 < 2\pi$. We draw a segment $T = T(a_1, a_2)$ and construct two congruent plane triangles $a'_1a'_2a'$, $a''_1a''_2a''$ such that

$$|a'_1 - a'_2| = |a''_1 - a''_2| = a_1a_2$$

$$\angle a'a'_1a'_2 = \angle a''a''_1a''_2 = \alpha_1/2, \ \angle a'a'_2a'_1 = \angle a''a''_2a''_1 = \alpha_2/2.$$

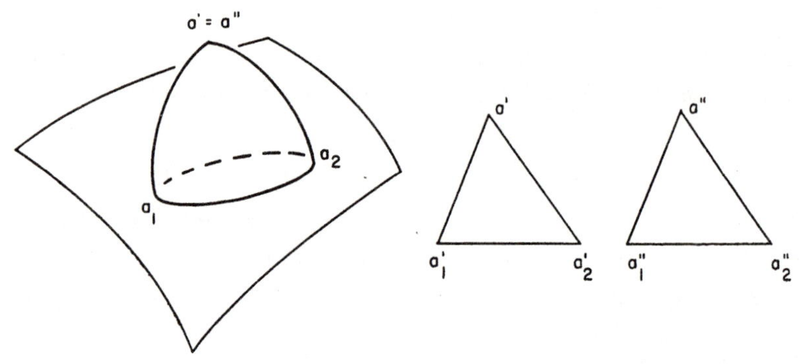

Figure 10

We cut the sphere with the metric ϱ_0 open along T and glue the two triangles to the two sides of T, and the triangles to each other such that the following vertices and edges are identified (=) $a_1 = a'_1 = a''_1$; $a_2 = a'_2 = a''_2$; $a' = a''$; $E(a', a'_1) = E(a'', a''_1)$; $E(a', a'_2) = E(a'', a''_2)$; one side of T and $E(a'_1, a'_2)$; the other side of T and $E(a'_1, a''_2)$.

This yields a $(c-1)$-Metric ϱ_1 on S^2. For the angles at a_1 and a_2 are now 2π, so that these points are no longer corners, but there is a new corner at $a' = a''$ whose curvature is

$$2\pi - 2[\pi - (\alpha_1 + \alpha_2)/2] > 0.$$

We deform ϱ_0 continuously into ϱ_1 by taking the points $x' = (1-t)a'_2 + ta'$, $x'' = (1-t)a''_2 + ta''$ and gluing the triangles $x'a'_1a'_2$ and $x''a''_1a''_2$ to the two sides of the cut T exactly as $a'a'_1a'_2$ and $a''a''_1a''_2$ above. For $0 < t < 1$ this yields a c-Metric ϱ_t with corners at $a_1 = a'_1 = a''_1$ and $x' = x''$ but no corner at $a_2 = a'_2 = a''_2$.

The case $c = 4$, $\alpha_1 = \alpha_2 = \alpha_3 = \alpha_4 = \pi$ is reduced to the general case by the *opposite process*:

There are two vertices $a_i \neq a_j$ such that $T(a_i, a_j)$ is unique. For if, for example, $T(a_1, a_3)$ is not unique then we know that two such segments divide S^2 into two domains, one of which contains a_2 and the other a_4. Then $T=T(a_1, a_2)$ is unique. Assume this to be the case, and for b sufficiently close to a_2 draw the (unique) segment $T(a_2, b)$ which forms with T (because of $\alpha_2 = \pi$) two angles of measure $\pi/2$. Then, for b close to a_2, there are two segments T' and T'' from a_1 to b close to T which tend for $b \rightarrow a_2$ to T, because $T(a_1, a_2)$ is unique. We cut the two geodesic triangles (close to T) bounded by T, $T(a_2, b)$ and T' or T'' respectively out and identify T' with T''. The resulting metric has corners at a_1 and b and is transformed continuously into the given metric by moving b continuously towards a_2. The curvature at a_1 has increased, but the curvature at b is less than π, for the total angle at b equals 2π minus twice the angle at b in the triangle cut out. These triangles are congruent to plane triangles, and since the angles at a_2 equal $\pi/2$, the angles at b are smaller than $\pi/2$.

The deformation ϱ_t of ϱ_0 is clearly such that there are for $t < 1$ proper nets N_t of the same structure which vary continuously with t. On ϱ_1 one vertex has disappeared, say, the vertex corresponding to a_1, but there is a definite point a'_1 which is the limit of the points a_1^t corresponding to a_1, and this gives rise to an improper net N_1 to which N_t converges for $t = 1$ together with the lengths of corresponding edges.

We now come to the *proof of* III. By hypothesis ϱ_1 is realizable by a polyhedron P_1. The point b_1 on P_1 corresponding to a_1' is an interior point of a face or an edge. We move b_1 into the exterior of P_1 to a position b (outside of the plane of P_1 if P_1 is degenerate). Then the convex closure P of P_1 and b is a polyhedron which tends to P for $b \rightarrow b_1$, hence the intrinsic metric ϱ_b of P will tend to ϱ_1. To apply I and II it must be verified that this is still true in terms of a net N_b close to N_1 and hence to N_t for t close to 1, see A., p. 260.

We want to show that ϱ_0 *can be "connected" to* ϱ_b *by c-Metrics.* The difficulty derives from the fact that a_1' is not a corner of ϱ_1, so that changing the lengths of the edges of N_1 may produce a corner with negative curvature. If $(r_1, \ldots, r_e) = r$ is the space of the lengths of the net edges and r^1 corresponds to N_1, then the nets with curvature 0 at a'_1 form, in a neighborhood of r^1, a hypersurface H through r^1, and the curvature at the point corresponding to a'_1 is positive on one side of H and negative on the other. For t close to 1 and b close to b_1, both lie close to ϱ_1 and on the same side of N, hence can in r-space be connected by a path not meeting H.

Thus we obtain a continuous family, of c-Metrics δ_t, $0 \leqq t \leqq 1$, connecting $\delta_0 = \varrho_0$ to a realizable c-Metric $\delta_1 = \varrho_b$. To apply I we must still convince ourselves that the δ_t can all be chosen non-degenerate. We may assume that δ_0 is non-degenerate because degenerate metrics are, by definition, realizable.

A degenerate P polyhedron is a closed plane convex polygon with vertices $a_1, \ldots, a_c$ covered twice. We construct a net on P by drawing on both sides the diagonals $T(a_1, a_i)$, $i = 2, \ldots, c$. Let $s_1, \ldots, s_{2c-6}$ be the lengths of these diagonals so numbered that $s_{2\nu-1}$ and $s_{2\nu}$ correspond to $a_{\nu+1}$; then

$$(17.6) \qquad s_1 = s_2,\ s_3 = s_4, \ldots, s_{2c-7} = s_{2c-6}.$$

It is rather easy to see that a metric with a net of the same structure close to this net is *non-degenerate when at least one of the above equations is not satisfied*, A., p. 264.

We now return to our last family δ_t, and denote by $\bar{t}$ the greatest lower bound of those t for which δ_t is degenerate. It follows from II that then $\delta_{\bar{t}}$ is degenerate, hence $\bar{t} > 0$. If P realizes $\delta_{\bar{t}}$, then we choose on P a net of the above type, and on all polyhedra close to P nets of the same structure close to this net. The equations (17.6) give $c - 3$ conditions for degeneracy.

If $c > 4$, then $e - (c-3) < e - 1$, hence the $s = (s_1, \ldots, s_e)$ corresponding to degenerate polyhedra do not decompose the s-space. This remains true, of course, for the points r in the space or edge lengths r_i corresponding to degenerate polyhedra in the net belonging to δ_t with t close to $\bar{t}$, so that for sufficiently small positive $\bar{t} - t_0$ the metric δ_{t_0} can be connected to any non-degenerate Metric δ close to $\delta_{\bar{t}}$ through non-degenerate c-Metrics ("connecting" refers, of course, to curves in r-space).

On the other hand we know that there are realizable non-degenerate c-Metrics δ arbitrarily close to a degenerate c-Metric. δ_t, $0 \leqq t \leqq t_0$, and a curve in r-space from δ_{t_0} to δ not meeting the degenerate locus, accomplishes step III if $e > 4$.

If $c = 4$ then $e - (c - 3) = e - 1$ and the degenerate locus is a hypersurface. It is quite easily seen that both sides of this hypersurface contain realizable non-degenerate c-Metrics, A., p. 265, and the conclusion is the same.

If we call the final family of c-Metrics again ϱ_t, $0 \leqq t \leqq 1$, as in III, then I and II clearly entail that the set of t corresponding to realizable c-Metrics is both open and closed in [0, 1], hence all ϱ_t, in particular ϱ_0, are realizable.

The method of passing continuously from a realizable metric to a given metric, and using I and II to show that the given metric is also realizable, was used by Weyl [1] in his, not complete, proof of the realizability of a metric with positive curvature on S^2.

18. Weyl's problem

The existence of a closed convex surface realizing a given intrinsic metric on the sphere is obtained from the realization of polyhedra by a limit process, but first we must find a characteristic property of the intrinsic metric of a convex surface. It was already mentioned that the *convexity condition* (13.5) *is characteristic, but not quite satisfactory*, because it is hard to prove and very strong; moreover, it deviates too much from the classical condition of non-negative curvature.

Instead of the convexity condition, Alexandrov requires that the *excess of a geodesic triangle be non-negative*. This presupposes a concept of angle independent of the convexity condition. In order to assume as little as possible, angle is defined as a lower limit which always exists. The weakness of the hypotheses makes the proof very long. The details are not very interesting and most readers would probably accept some of the intermediate results, like the existence of the angle (13.2) for segments, as hypotheses. We therefore omit the proofs for such results and refer the reader to A.

Consider a two-dimensional manifold with an intrinsic metric. We know from (10.11) that every point has a neighborhood, any two points of which can be connected by a segment. From Section 13 we remember the definition of the angle $p(x, y)$ for three points p, x, y with $p \neq x$, $p \neq y$ by

$$\cos p(x, y) = \frac{px^2 + py^2 - xy^2}{2px \cdot py}, \quad 0 \leqq p(x, y) \leqq \pi.$$

As in (13.8) let $S : x(s),\ 0 \leqq s \leqq \alpha;\ T : y(t),\ 0 \leqq t \leqq \beta,$ $x(0) = y(0) = p$ be two segments emanating from p, where s and t are arc length, and put $p(x(s), y(t)) = p(s, t)$. The angle $\lim_{s\to 0+,\, t\to 0+} p(s, t)$ will in general not exist, and even if it exists, it need not exist in the strong sense (13.12). We therefore put

$$(18.1) \qquad \delta'(S, T) = \liminf p(s_\nu, t_\nu)$$

where $s_\nu \cdot t_\nu \to 0$ and a segment $T(x(s_\nu), y(t_\nu))$ exists whose upper closed limit [6] lies on $S \cup T$. If both $s_\nu \to 0$ and $t_\nu \to 0$ then $T(x(s_\nu), y(t_\nu))$ tends to p and the latter condition is automatically satisfied. Under the convexity condition $\lim p(s_\nu, t_\nu)$ exists, see (13.12).

The angle $\delta'(S, T)$ is not yet our final concept. Since s_ν and t_ν do not necessarily both tend to zero, $\delta'(S, T)$ will depend on S and T as a whole. If S, T are replaced by subsegments S_0, T_0 then, in general, $\delta'(S_0, T_0) > \delta'(S, T)$. Therefore we define

$$(18.2) \qquad \delta(S, T) = \inf \delta'(S', T')$$

where S' and T' traverse all segments which have with S and T respectively, proper segments with origin p in common.

A *convex triangle* is a convex set, see Section 14, bounded by three segments $C = T(a, b)$, $A = T(b, c)$, $B = T(c, a)$. The excess of this triangle is the sum of the angles $\delta(A, B)$, $\delta(B, C)$ $\delta(C, A)$ minus π. *A metric with non-negative curvature is an intrinsic metric on a two-dimensional manifold where every point has a neighborhood such that every convex triangle in it has non-negative*

[6] In Hausdorff's sense, i.e., the set of all accumulation points of sequences $\{q_\nu\}$, $q_\nu \in T(x(s_\nu), y(t_\nu))$.

excess. We know that the local convexity condition implies non-negative curvature, see (13.13).

The fundamental realization theorem is:

(18.3) THEOREM. *A metric on S^2 with non-negative curvature can be realized as a closed convex surface in E^3.*

Here we must again admit doubly covered plane convex domains. By Pogorelov's general monotypy theorem, the realization is unique, this fact and (18.3) may be briefly formulated as:

(18.4) *An intrinsic metric on S^2 with non-negative curvature is monotypic.*

The definition (18.2) of angle is not very satisfactory. The *upper angle*

$$\delta''(S, T) = \limsup_{s\to 0+,\, t\to 0+} p(s, t)$$

behaves better because

$$\delta''(S, T) = \lim_{s\to 0+} p(s, t)$$

see A. [9], so that we need not consider the complicated limits (18.1) and (18.2). However it does not suffice to postulate that the excess be non-negative to characterize convex surfaces. In any two-dimensional Minkowski geometry (see for example Busemann [3], Section 17) the excess is non-negative. This observation stopped Alexandrov in his book from using $\delta''(S, T)$. Actually the development of a theory of general (i.e., not necessarily convex) surfaces, about which we will briefly report in the Conclusion, has shown that this approach is *preferable.* Only, the requirement of non-negative excess must be supplemented by the boundedness of the total curvature: the sum of the excesses of any finite number of non-overlapping convex triangles does not exceed a finite number M. Of course, it turns out that then $M = 4\pi$ will do. [7]

The following facts, which we know as consequences of the

[7] We do not follow this procedure because a detailed treatment does not seem to exist in the literature. Carrying the details out here would have required too much space.

convexity condition are deduced in A., pp. 274–293 from the hypothesis of non-negative curvature.

The ordinary angle (13.2) *exists for any two sides of a convex triangle and equals the angle* (18.2), *at least when* the sides are not prolongations of each other. (13.13) holds for small convex triangles, i.e., *each angle is at least as large as the corresponding angle in the euclidean triangle with the same sides.* The *inclusion property* (13.10) holds. Finally, the *complete angle exists and is at most* 2π *at any point* p *which is origin of a finite number of segments* $T_1, \ldots, T_n$, *in cyclical order, such that* T_i *and* T_{i+1} *bound a convex sector* $i = 1, \ldots, n$, $T_{n+1} = T_1$. This means that $\Sigma\delta(T_i, T_{i+1})$ is independent of the choice of the T_i, as long as the sectors are convex, and is at most 2π.

The lower-semicontinuity of angles (13.14) applied to $C_i = C$ becomes a theorem on the intrinsic metric of C. This theorem holds in our case and entails together with the additivity (13.4): *if a segment* T *emanates from the interior point of the segment* S, *decomposing* S *into the segments* S_1 *and* S_2, *then* $\delta(S_1, T)+\delta(T, S_2)=\pi$, see A., pp. 147–148. Therefore, each of the two sector angles $\delta(S_1, S_2)$ equals π.

The inclusion property entails, see Section 14, that each point lies in a convex polygonal domain homeomorphic to a disk and with arbitrarily small diameter. Moreover *the triangulation theorem* (14.2) *holds.* In particular, a convex polygonal domain $\bar{D}$ can be triangulated into arbitrarily small proper convex triangles.

A vertex of such a triangulation which lies in the interior D of $\bar{D}$ satisfies the hypotheses for the existence of a complete angle $\leqq 2\pi$. Therefore we can apply (14.3) to D. The *excess* $\epsilon(D)$ of D is defined by the left side of (14.3), but is independent of the triangulation of $\bar{D}$ into convex triangles.

Consider a triangulation of $\bar{D}$ into proper convex triangles $T_1, \ldots, T_t$ of diameter less than δ, $\delta > 0$. For each T_i we construct a euclidean triangle T'_i with sides of the same lengths, and then we identify the T'_i in exactly the same way as the T_i. This yields an abstract polygonal manifold $\bar{D}'$, with a boundary if $\bar{D}$ has one, whose interior D' corresponds to D. Then D' is a polyhedral metric with non-positive curvature (see last section), because the complete angle at any vertex in D does not exceed 2π and the angles in any T_i are at most as large as the corresponding angles in T_i. Since $\bar{D}$ is convex the angles measured in D at the boundary of $\bar{D}$ are at most π, therefore the same holds at the boundary of $\bar{D}'$, so that a shortest connection in $\bar{D}'$ of two points in D' lies entirely in D'.

In order to *compare the metrics* of $\bar{D}$ and $\bar{D}'$, we map $\bar{D}$ on $\bar{D}'$ as follows: a side of any T_i is mapped isometrically on the corresponding side of T_i', and this isometric mapping of the boundary of T_i on that of T_i' is continued to a homeomorphic mapping of the interior of T_i on the interior of T_i'. This yields a homeomorphism $x \to x'$ of $\bar{D}$ on $\bar{D}'$. We denote the intrinsic distance in $\bar{D}$ of two points x, y in $\bar{D}$ by $\varrho(x, y)$ and that for two points x', y' in $\bar{D}'$ by $\varrho'(x', y')$.

We want to show that, for small δ, the distances $\varrho(x, y)$ and $\varrho'(x', y')$ of corresponding pairs of points differ little. This is comparatively simple in one direction:

$$\varrho'(x' y') < \varrho(x, y) + (\epsilon(D) + 2)\delta. \tag{18.4}$$

The proof rests on an interesting estimate which is obtained by very elementary means:

(18.5) *Let abc be a convex triangle in a manifold with non-negative curvature, and a'b'c' an euclidean triangle with the same sides. If x and y lie on the sides $T(a, b)$ and $T(a, c)$ of abc and x', y' are the corresponding points* (i.e., $ax = |a' - x'|, \ldots$) *on $E(a', b')$ and $E(a', c')$ then*

$$\left| xy - |x' - y'| \right| \leqq 4 \ (ax \cdot ay)^{1/2} \sin [\epsilon(abc)/4].$$

We put

$$s = ax = |a' - x'|,\ t = ay = |a' - y'|,\ u = xy,\ u' = |x' - y'|,$$
$$2\alpha = a(x, y),\ 2\alpha' = a'(x', y').$$

The definition of $a(x, y)$, $a'(x', y')$ and the double-angle formulas for cosine yield:

$u^2 = (s - t)^2 + 4st \sin^2 \alpha, \quad u'^2 = (s - t)^2 + 4st \sin^2 \alpha',$ whence $u + u' \geqq 2(st)^{1/2} (\sin \alpha + \sin \alpha'), \quad u^2 - u'^2 = 4st (\sin^2 \alpha - \sin^2 \alpha')$

and

$$(18.6) \quad u - u' \leqq 2(st)^{1/2} |\sin \alpha - \sin \alpha'| \leqq 4(st)^{1/2} \sin (|\alpha - \alpha'|/2).$$

The sum of the angles in $a'b'c'$ is π, and each angle in abc is at least as large as the corresponding angle in $a'b'c'$. Therefore, if 2α is the angle at a in abc since $2\alpha'$ is the angle at a' in $a'b'c'$

$$0 \leqq 2\bar{\alpha} - 2\alpha' \leqq \epsilon(abc), \text{ or } 2\bar{\alpha} - \epsilon(abc) \leqq 2\alpha' \leqq 2\bar{\alpha}.$$

Because abc is convex there is a segment $T(x, y)$ in abc which cuts a convex triangle axy off abc. The same argument and the fact that the curvature is non-negative yield

$$2\bar{\alpha} - \epsilon(abc) \leqq 2\bar{\alpha} - \epsilon(axy) \leqq 2\alpha \leqq 2\bar{\alpha}$$

and together with the preceding inequality

$$|2\alpha - 2\alpha'| \leqq \epsilon(abc).$$

Because $\epsilon(abc) < 2\pi$

$$\sin (|\alpha - \alpha'|/2) \leqq \sin [\epsilon(abc)/4].$$

Substituting this into (18.6) proves (18.5).

Returning to the assumption of (18.4), each T_i has diameter less than δ, hence (18.5) gives for two points x, y on the boundary of any T_i and the two corresponding points of T'_i

$$(18.7) \quad \varrho'(x', y') \leqq \varrho(x, y) + 4\delta \sin \epsilon(T_i)/4 \leqq \varrho(x, y) + \delta\epsilon(T_i).$$

A shortest join L of two points x, y in $\bar{D}$ has length xy because $\bar{D}$ is convex. The intersection of L with any T_i is either empty, or a point, or a proper segment. For, because T_i is convex, L could be shortened if the intersection had any other shape. Thus there are points $x_0 = x, x_1, \ldots, x_n = y$ in this order on L such that any two consecutive x_i lie in one T_i and x_{k-1}, x_k lie for $2 \leqq k \leqq n - 1$ on the boundary of the same T_{i_k} and the T_{i_k} are all different. Now

$$\varrho(x, y) = \text{length } L = \sum_{k=1}^{n} \varrho(x_{k-1}, x_k), \quad \varrho'(x', y') \leqq \sum_{k=1}^{n} \varrho'(x'_{k-1}, x'_k).$$

Because a side of any T_i is smaller than δ, the same holds for T'_i, hence $\varrho'(x', x'_1) < \delta$, $\varrho'(x'_{n-1}, y') < \delta$ and by (18.7)

$$\varrho'(x'_{k-1}, x'_k) \leqq \varrho(x_{k-1}, x_k) + \delta\epsilon(T_{i_k}), \quad 2 \leqq k \leqq n-1.$$

Because the T_{i_k} are distinct, adding these inequalities yields the desired result:

$$\varrho'(x', y') < 2\delta + \sum_{k=2}^{n-1} \varrho(x_{k-1}, x_k) + \delta \sum_{k=2}^{n-1} \epsilon(T_{i_k}) < 2\delta + \varrho(x, y) + \delta\epsilon(D).$$

The analogous inequality

$$\varrho(x, y) < \varrho'(x', y') + c\delta \tag{18.8}$$

where c is an absolute constant (in (18.4) we may replace $\epsilon(D)$ by 4π) *is correct for convex polygonal domains* D with $\epsilon(D) < \pi$, but cannot be proved by simply reversing the rôles of ϱ and ϱ' in the preceding argument, because we do not know that a T'_i is a convex set for ϱ' and is therefore crossed at most once by a shortest join for ϱ'. Actually the proof requires many case distinctions and is involved, see A., pp. 295–298 and pp. 302–306.

To *prove theorem* (18.3) take a sequence $\delta_1 > \delta_2 > \ldots, \delta_i \to 0$. The sphere with the given metric $\varrho(x, y)$ of non-negative curvature is a convex polygonal domain. For each i we triangulate S^2 into proper convex triangles of diameter less than δ_i and construct a polyhedral metric $\varrho_i(x, y)$ as above $\varrho'(x', y')$, only we now interpret x and y as points of the same sphere S^2. By (17.1) each $\varrho_i(x, y)$ can be realized as a closed convex polyhedron P_i which passes through a fixed point z. For any point x on S^2 we denote by x^i the point corresponding to x on P_i. Then by (18.4)

$$|x^i - y^i| \leqq \varrho_i(x, y) \leqq \varrho(x, y) + \beta\delta, \ \beta = 4\pi + 2, \tag{18.9}$$

so that $\cup P_i$ is a bounded set. By Blaschke's selection theorem, see K., p. 34, we extract from $\{P_i\}$ a subsequence $\{P_k\}$ which converges (always in Hausdorff's sense) to the boundary C of a convex body which, *a priori*, may have dimension 3, 2, 1, or 0. We wish to show that C realizes $\varrho(x, y)$. Dimension 2 corresponds to our doubly covered plane convex sets and is admitted; dimension 1 or 0 would yield a segment or point and must be ruled out.

The union of all vertices occurring in the triangulations yielding the ϱ_i, $i = 1, 2, \ldots$ form a countable set A on S^2. The diagonal process furnishes a subsequence $\{P_n\}$ of $\{P_k\}$ such that the point $a_n \epsilon P_n$ corresponding to a point $a \epsilon A$, converges for each $a \epsilon A$. Then x_n converges for each $x \epsilon S^2$. For x has distance at most δ_n from a suitable $a \epsilon A$ and by (18.9)

$$|x_n - x_m| \leqq |x_n - a_n| + |a_n - a_m| + |a_m - x_m|$$
$$\leqq \delta_n + \beta\delta_n + |a_n - a_m| + \delta_m + \beta\delta_m.$$

Because a_n converges and $\delta_n \to 0$, the sequence $\{x_n\}$ converges. We put

$$\lim x_n = \bar{x} \epsilon C, \ x \epsilon S^2, \ x_n \epsilon P_n.$$

When C is a doubly covered plane domain an ambiguity arises: we have to ascertain that x_n converges to a "definite one" of the two points of C corresponding to one. It is intuitively clear that this can be accomplished by passing to a subsequence of $\{P_n\}$, for which two points of P_n corresponding to a point of C appearing once on each side of C converge individually. The explicit argument is found in A., pp. 311–312.

We must show that $\varrho(x, y) = (\bar{x}\bar{y}, C)$ and know from (11.3) that

$$\varrho_n(x, y) = (x_n y_n, P_n) \to (\bar{x}\bar{y}, C).$$

Strictly speaking, we did not prove (11.3) when C degenerates into a segment or a point, but it is obviously also correct in this case. (18.9) yields at once $(\bar{x}\bar{y}, C) \leqq \varrho(x, y)$.

Since (18.8) is established only for convex polygonal domains D homeomorphic to a disk with $\epsilon(D) < \pi$, the argument for $(\bar{x}\bar{y}, C) \geqq \varrho(x, y)$ is slightly more involved.

Consider the triangulation leading to P_n. Each triangle is a convex polygonal domain homeomorphic to a disk. Each T_i goes under $x \to \bar{x}$ into a set $\bar{T}_i$ on C. If $\epsilon(T_i) < \pi$ we can apply (18.8) to T_i. Take a shortest connection $\bar{L}$ from $\bar{x}$ to $\bar{y}$ on C and on $\bar{L}$ a maximal subarc, say from $\bar{a}$ to $\bar{b}$, which lies in such a $\bar{T}_i$. Then we know

$$\varrho(a, b) = \lim \varrho_n(a_n, b_n) = (\bar{a}\bar{b}, C).$$

We associate with a and b a segment $T(a, b)$ for the metric ϱ in S^2 and do the same for all the maximal subarcs of $\bar{L}$ in $\bar{T}_i$, and obtain a set of segments for ϱ, the sum of whose lengths equals the length of the arcs of $\bar{L}$ in $\bar{T}$.

We do the same for all T_i with $\epsilon(T_i) < \pi$ avoiding duplication. If $\bar{L}$ contains a point of a $\bar{T}_i$ with $\epsilon(T_i) \geqq \pi$ we associate to the arc of L from its first common point c with $\bar{T}_i$ to its last d, a segment $T(c, d)$. Then $\varrho(c, d) < \delta_n$. These segments, for the metric ϱ, form a curve W from x to y. There are at most four T_i with $\epsilon(T_i) \geqq \pi$ because $\epsilon(S^2) = 4\pi$. Consequently, using λ to denote length of a curve,

$$\varrho(x, y) \leqq \lambda(W) \leqq \lambda(L) + 4\delta_n = (\bar{x}\bar{y}, C) + 4\delta_n.$$

This proves $\varrho(x, y) = (\bar{x}\bar{y}, C)$. Because ϱ is a metrization of S^2 it is obvious that C cannot be a segment or a point. *This completes the proof of* (18.3).

The heading of this section is "*Weyl's problem,*" *which in its original form is this*:

Given an analytic line element $ds^2 = E(u, v)du^2 + 2F(u, v)\,du\,dv + G(u, v)dv^2$ on S^2 with $EG - F^2 > 0$ and positive curvature, does there exist an analytic surface $\mathfrak{x}(u, v) = (x_1(u, v), x_2(u, v), x_3(u, v))$ in E^3 such that

$$\mathfrak{x}_u^2 = E, \ \mathfrak{x}_u \cdot \mathfrak{x}_v = F, \ \mathfrak{x}_v^2 = G? \tag{18.10}$$

The analyticity of E, F, G, x_1, x_2, x_3 means that these functions are analytic as functions of local analytic parameters u, v on S^2; compare Section 5.

Weyl [1] posed this problem; it was solved partially by him and completely by Lewy [1]. Because the realization is unique, see (18.4), this implies that the realization obtained from (18.3) is analytic. On the basis of our present results we can only say that it is of class C', see (5.2) (it therefore realizes the line element in the sense that (18.10) holds). However, this is not the moment to discuss the question. For we will see presently that (18.3), just because it is not restricted to smooth metrics, furnishes realizations of metrics with non-negative curvature, defined on suitable subsets of the sphere, by non-complete convex surfaces, and these realizations are, in general, by no means unique. We will, therefore, have to answer the question, *whether any, not necessarily complete, convex surface with an analytic line element is analytic*, for example, a surface intrinsically isometric to a subset of an ellipsoid.

19. Local realization of metrics with non-negative curvature

The proof of theorem (18.3) required a considerable effort, for it is based not only on Section 18, but also on Sections 13, 14, 16, 17, and yet many details were omitted. This effort is justified because (18.3) does not only solve Weyl's Problem in a form which comprises the most general closed convex surface, but has some truly astonishing implications in the theory of deformation of smooth surfaces, which transcend anything attainable by the classical methods. A simple example will illustrate this.

We know that a manifold M with non-negative curvature can be realized as a convex surface in E^3 if M is homeomorphic to S^2. That the corresponding statement cannot be correct for general M is clear from the example of the projective plane with the elliptic (or spherical) metric. But we expect:

(19.1) *Every point of a manifold M with non-negative curvature has a neighborhood which can be realized as a (non-complete) convex surface in E^3.*

We deduced in Section 14 from the inclusion property that every point of M lies inside a convex polygonal domain homeomorphic to a disk, hence (19.1) will follow from:

(19.2) THEOREM. *If $\bar{D}$ is a convex polygonal domain homeomorphic to a disk on a manifold with non-negative curvature, then the interior D of $\bar{D}$ can be realized as a convex surface in E^3.*

In this section it will prove convenient to use the term simple polygon on a manifold with an intrinsic metric for a curve either of the form

$$(19.3)\qquad \begin{aligned} &\textit{closed:} \quad \sum_{i=1}^{n} T(x_i, x_{i+1}),\ x_{n+1} = x_1,\ n > 3, \text{ or} \\ &\textit{open:} \quad \sum_{i=-\infty}^{\infty} T(x_i, x_{i+1}), \text{ and if } p_i \in T(x_i, x_{i+1}) \text{ then } \{p_i\} \end{aligned}$$

does not have an accumulation point, moreover, in both cases $T(x_{i-1}, x_i) \cap T(x_j, x_{j+1}) = 0$ for $i < j$ and $= x_i$ for $i = j$.

The boundary of D in (19.2) can (in many ways) be written as a simple closed polygon of the above type such that each $x_i x_{i+1}$ is smaller than the sum of the remaining $x_j x_{j+1}$. We construct a closed convex polygon $x'_1, x'_2, \ldots, x'_n, x'_{n+1} = x'_1$ in E^2 with $|x'_i - x'_{i+1}| = x_i x_{i+1}$ bounding in E^2 a convex set $\bar{D}'$ with a non-empty interior D'. We identify the boundaries of $\bar{D}$ and $\bar{D}'$, by mapping x_i on x'_i and $T(x_i, x_{i+1})$ isometrically on $E(x'_i, x'_{i+1})$. Then the metrics on $\bar{D}$ and $\bar{D}'$ induce in a natural way (see below) an intrinsic metric on $\bar{D} \cup \bar{D}'$ which is homeomorphic to S^2. Since the angles at x_i of $\bar{D}$ measured in D and at x'_i of $\bar{D}'$ measured in D' are at most π, it is plausible, and will be shown presently under more general conditions, that the metric of $\bar{D} \cup \bar{D}'$ has non-negative curvature. By (18.3) $\bar{D} \cup \bar{D}'$ can be realized as a closed convex surface in E^3, and the part of this surface corresponding to D realizes D.

Since there is much arbitrariness in the way the representation of the boundary of $\bar{D}$ as a simple polygon and the angles in $\bar{D}'$ are for $n > 3$ largely arbitrary, *this construction furnishes not only one, but many different realizations of* D.

Alexandrov says that the surface $\bar{D} \cup \bar{D}'$ or its realization in E^3 originates from D and D' by *"gluing."* This process is very important, therefore we formulate it in a general fashion:

Consider two-dimensional manifolds $M_1, \ldots, M_n$ with intrinsic metrics $\varrho_1, \ldots, \varrho_n$. On each M_i take a set $\bar{D}_i$ which is the closure of an open connected set D_i and whose boundary consists of a finite number of disjoint curves $C_1^i, \ldots, C_{k_i}^i$, where C_j^i is (with the metric ϱ_i) either a closed Jordan curve, or a topological image of the real axis. We assume that each subarc of a C_j^i is rectifiable. A section of a C_j^i is either all of C_j^i or a subarc consisting of more than one point, or if C_j^i is open also a point of C_j^i and all points on C_j^i on one side of it.

The D_i are *glued* together by identifying the C_j^i according to the following rules:

1) *The C_j^i are decomposed into a finite number of sections. These sections are pairwise identified such that any two indentified subarcs of identified sections have the same length.*

2) *After this identification $\bar{D}_1 \cup \ldots \cup \bar{D}_n$ is a topological manifold M.*

It follows that an interior point of two identified sections has a neighborhood which is divided by this section into two sets homeomorphic to half planes, and each of these sets lies entirely in one $\bar{D}_i$. These two $\bar{D}_i$ need not be distinct: let $\bar{D}_1$ be a closed hemisphere bounded by the great circle B. Two diametrically opposite points p_1, p_2 of B divide B into two semi great circles. These may be identified by associating points on B with equal distance from p_1. The manifold obtained is, by the way, easily seen to have non-negative curvature. In cases like this we can, of course, dissect $\bar{D}_1$ into two domains, then the two $\bar{D}_i$ will be distinct.

The origin p of identified sections will, in general, belong to several $\bar{D}_i$. Because of 2) these can be arranged in cyclical order.

The metrics $\varrho_1, \ldots, \varrho_n$ induce on M a natural intrinsic metric: for two points x, y of M we consider all arcs C_{xy} from x to y such that C_{xy} is the union of a finite number of non-overlapping subarcs, each of which lies in a $\bar{D}_i$, and hence has a length as curve in M_i. We define $\lambda(C_{xy})$ as the sum of the lengths of these arcs and put $xy = \inf_{C_{xy}} \lambda(C_{xy})$. Although we did not admit all curves from x to y it is quite clear that xy is an intrinsic metric in M. We say that *M with this metric is the result of gluing $\bar{D}_1, \ldots, \bar{D}_n$ together.*

We apply this idea first to the case where the C_j^i are simple polygons of the type (19.3). If $M_1, \ldots, M_n$ have non-negative curvature then M can have non-negative curvature, only if at any point p not in $\cup D_i$ the sum of the sector angles at p measured in D_i of the boundaries of those $\bar{D}_i$ which contain

p does not exceed 2π. This is also sufficient. However, before (19.1) has been established, we know this and the relation of the sector angle to the angle (13.2) only when the $\bar{D}_i$ are convex polygonal domains. The $\bar{D}$ and $\bar{D}'$ entering the proof of (19.2) are of this type. Therefore *we prove that M has non-negative curvature under the hypothesis that these relations prevail. This implies* (19.1), *which allows us to eliminate that hypothesis.*

It is clear that every convex triangle in a suitable neighborhood of a $p \subset \cup D_i$ has non-negative excess. If p lies on the boundary of some D_i, then there are curves $T_1, \ldots, T_m, T_{m+1} = T_1$ issuing from p with the following properties: $T_i \cap T_j = p$; the T_i follow in cyclical order, T_i and T_{i+1} lie on the boundary of the same D_j which, for simplicity of notation, we assume to be D_i, moreover T_i is a segment in both M_{i-1} and M_i, $i = 2, \ldots, n$ and T_n in M_n and M_1.

A segment S in M from p to a point of D_i (everything takes place in a sufficiently small neighborhood of p) lies, except for p, in D_i. For if S contained a point $q \in D_j$, $j \neq i$, we could take the first points q_1 and q_2 proceeding from q on S in either direction, which lie on the boundary of D_j. If p lies on $T_{j-1} \cup T_j$ between q_1 and q_2, then S could be shortened. Hence q_1 and q_2 lie both on T_{j-1} or both on T_j. Since T_{j-1} and T_j are segments in M_{j-1}, which satisfies the inclusion property, the subarc of T_{j-1} or T_j from q_1 to q_2 is the only segment $T(q_1, q_2)$ in M_j, hence S could be shortened by replacing the arc from q_1 to q_2 by $T(q_1, q_2)$. Therefore S lies in D_i. The inclusion property for M_i shows that S cannot contain a point of $T_i \cup T_{i+1}$ other than p. This result yields easily that *M has the inclusion property.*

Consider now a convex triangle Δ with one vertex at p. If T' and T'' are the sides of Δ issuing from p and no T_i lies (initially) inside Δ, then Δ lies in one D_i and the ordinary angle exists for T' and T''. In the cyclical order $T_0 = T'$, $T_1, T_2, \ldots, T_{m-1}$, $T_n = T''$ let $T_1, T_2, \ldots, T_{m-1}$ lie between T' and T'' and in Δ. Then the ordinary angles between T_i and T_{i+1} exist. Because Δ is convex it follows from (13.4) that the angle between T' and T'' exists. If T' and T'' are continuations of each other, then the ordinary angle between T' and T'' evidently exists and equals π.

The concept (18.2) of the angle $\delta(T', T'')$ entering the definition of non-negative curvature is not the ordinary angle even if it exists. However, it can be shown that the angle exists in the strong sense (13.12), therefore the ordinary angle coincides with $\delta(T', T'')$. The proof of (13.12) in A., pp. 134–138 is so arranged that it applies to the present case, see A., p. 319.

It remains to be shown that $\epsilon(\Delta) > 0$ for any convex triangle Δ in a neighborhood of p. This is obvious if lies in one D_i. If this is not the case, then the intersection of Δ with any D_i consists, because of the inclusion property, of a finite number of components. For the proof of (19.1) we observe that these components are convex when the D_i are convex polygonal domains, because they are intersections of such domains, see Section 14. By hypothesis the complete angle about any of the vertices of these components is at most 2π. Therefore the definition (14.4) of excess of a polygonal domain, the additivity of excess, and the fact that the M_i have non-negative curvature show that $\epsilon(\Delta) \geqq 0$.

Using the previous notations we have proved:

(19.4) THEOREM. *Gluing theorem for Polygonal Domains: Let* $M_1, \ldots, M_n$ *have non-negative curvature, and let the* C_j^i *be simple polygons of the type* (19.3). *Then the manifold originating by gluing the sets* $\bar{D}_i \subset M_i$ *together has non-negative curvature if, and only if, for any point* p *not in* $\cup D_i$ *the sum of the angles at* p *(measured in* D_i*) of the boundaries of those* $\bar{D}_i$ *that contain* p *does not exceed* 2π.

20. Existence of open surfaces. The general gluing theorem

Theorems (19.4) and (18.3) can be used to establish a result analogous to (18.3) for open surfaces:

(20.1) THEOREM. *A finitely compact metric with non-negative curvature in* E^2 *can be realized as an open convex surface in* E^3.

Here we have monotypy only when the total curvature $\nu(E^2)$ equals 2π, see Section 22.

Let $\varrho(x, y)$ be the given metric in E^2 and denote an auxiliary euclidean metrization of E^2 by $e(x, y)$ (the notation $|x - y|$ will be reserved for the realization space E^3). The finite compactness of $\varrho(x, y)$ entails that the relations $\varrho(p, x_\nu) \to \infty$ and $e(p, x_\nu) \to \infty$ are equivalent.

There is an increasing sequence $Q_1 \subset Q_2 \subset \ldots$ of polygonal domains (with respect to ϱ) with $\cup Q_i = E^2$. Such Q_i may be constructed as follows: On the circle $e(p, x) = 2i$, where p is any fixed point, we take points $q_1, q_2, \ldots, q_n, q_{n+1} = q$ in cyclical order so close together that $T(q_i, q_{i+1})$ lies in the circle $e(q_i, x) < 1$. Then the, not necessarily simple, polygon $\Sigma_{i=1}^n T(q_i, q_{i+1})$ lies in the ring

$$2i - 1 < e(p, x) < 2i + 1 \tag{20.2}$$

and decomposes E^2, because of the inclusion property, into a finite number of open sets, exactly one of which is unbounded. The boundary of this set is a simple polygon B'_i and a subset of $\cup T(q_i, q_{i+1})$. As Q_i we choose the set bounded by B'_i. The relation $Q_i \subset Q_{i+1}$ holds because B'_i lies in (20.2).

Through an arbitrary vertex v_i of Q_i we take a simple closed curve B_i of smallest length (for ϱ) such that Q_i lies inside or on B'_i. Such a B_i exists: in the first place B'_i is one of the curves "admissible" for competition in the minimum problem. Because $\lim_{\nu\to\infty} e(v_i, x_\nu) = \infty$ implies $\lim_{\nu\to\infty} \varrho(v_i, x_\nu) = \infty$ we have $\varrho(v_i, x) > \lambda(B'_i)$ on a circle $e(v_i, x) = \beta_i$ with a suitable β_i. If L_n is a sequence of simple closed curves passing through v_i and containing Q_i with

$$\lambda(L_n) \to \delta_i = \text{infimum of all admissible curves} \leqq \lambda(B'_i),$$

then L_n stays, at least from a certain n on, in $e(v_i, x) < \beta_i$. By (10.5) a suitable subsequence of $\{L_n\}$ converges to a curve B_i with $\lambda(B_i) \leqq \delta_i$. If B_i were not simple a suitable subcurve would contain Q_i and would be shorter, hence $\lambda(B_i) = \delta_i$.

Any point of B_i which does not lie on Q_i must be interior point of a segment on B_i. The same holds for points of B_i which are interior points of a side of Q_i. Because of the inclusion property B_i must then contain the whole side. Thus B_i *is a geodesic polygon whose only true vertices are vertices of* Q_i. *For any such vertex, with the exception of* v_i, *the angle of* B_i *measured in the exterior of* B_i *cannot be smaller than* π because B_i could then be shortened without entering the interior of Q_i.

Therefore the angles of all vertices of B_i with the possible exception of v_i, measured in the interior D_i of B_i are smaller than π. Let w_i be that point on B_i for which the two arcs A_{i1}, A_{i2} of B_i from v_i to w_i have length $\delta_i/2$. We identify A_{i1} and A_{i2} by associating points $x_k \in A_{ik}$, $k = 1, 2$, for which the subarcs of A_{ik} from v_i to x_k have equal length. Since the angle of B_i at v_i in D_i is less than 2π and the angles at the vertices of B_i in D_i are less than π, this identification satisfies the hypothesis of (19.4): gluing $\bar{D}_i = B_i \cup D_i$ to itself along A_{i1} and A_{i2} produces a manifold with non-negative curvature, which in this case is homeomorphic to S^2.

By (18.3) it can be realized as a convex surface K_i in E^3. For a given point $x \in E^2$ there is an i_x such that some $T(p, x)$ lies in D_i for $i \geqq i_x$. Therefore the image x_i of x on K_i is defined for $i \geqq i_x$ and K_i carries a curve of length $\varrho(p, x)$ from p_i to x_i. The point p_i exists for all i and we place K_i so that the point p_i falls into a fixed point p. Then

$$(20.3) \qquad |x_i - p| \leqq (x_i p_i, K_i) \leqq \varrho(x, p) \text{ for } i \geqq i_x.$$

Proceeding as in the proof of (18.3) we find a subsequence $\{K_k\}$ of $\{K_i\}$ which tends to the boundary K of convex set in E^3.[8]

A priori, K may be a closed or open convex surface (including doubly covered plane convex domains), the union of two parallel planes, a cylinder, a line, or a convex subset of a line. However, if we can show that K realizes $\varrho(x, y)$, then all cases, but an open convex surface, are automatically excluded, because $\varrho(x, y)$ is a finitely compact metrization of E^2.

As in Section 18 we take a countable dense subset A of E^2. By (20.3) each sequence $\{a_i\}$, $a \in A$, is bounded, hence we can find

[8] Blaschke's Selection theorem is applied successively for $\nu = 1, 2, 3, \ldots$ to the intersections of $\{K_i\}$ with the sphere $|p - x| \leqq \nu$. The diagonal sequence of these sequences is $\{K_k\}$.

a subsequence $\{K_n\}$ of $\{K_k\}$ such that $\{a_n\}$ converges for each $a \in A$. Then we conclude from (20.3), as in Section 18, that

$$\lim x_n = \bar{x} \text{ exists for each } x \in E^2.$$

We have to show that

$$(\bar{x}\bar{y}, K) = \varrho(x, y), \text{ for given } x, y \in E^2.$$

But this is obvious: A segment from x to y for the metric in $\bar{D}_n$ after the identification, or from x_n to y_n on K_n, can have length less than $\varrho(x, y)$ only when it contains a point of the identified arcs A_{n1} and A_{n2}. Therefore, as soon as the distances $\varrho(B_n, x)$ and $\varrho(B_n, y)$ exceed $\varrho(x, y)$, a segment $T(x, y)$ for ϱ remains a segment after the identification, or

$$(x_n y_n, K_n) = \varrho(x, y) \text{ for large } n,$$

and $(x_n y_n, K_n) \to (xy, \bar{K})$ by (11.3).

Theorem (19.4) is but a special case of a general gluing theorem. Assume the manifold M originates by gluing from manifolds $M, \ldots, M_n$ which are smooth convex surfaces, and that the sections of the C_j^i which we identified are smooth curves. We expect M to have *non-negative curvature if the following conditions are satisfied*: 1) *if p is an interior point of two identified sections belonging to C_r^i and C_s^j, then the sum of the geodesic curvatures at p of C_r^i towards $\bar{D}_i$ and of C_s^j towards $\bar{D}_j$ must be non-negative.* 2) *if p belongs to more than two $\bar{D}_i$ then the complete angle of M at p must not exceed 2π.* This is, indeed, correct and is in many respects the most interesting case.

To formulate a corresponding general theorem we remember from Section 15 that the right and left swerves exist for any Jordan arc A which has definite directions at its end points. If A has definite directions (towards either side) at every point, then the right swerve $s_r(A')$ is defined for every subarc A' of A. We say that the *right swerve has bounded variation on A*, if A has definite directions at each point and if with a suitable $M(A)$

$$|s_r(A_1)| + \ldots + |s_r(A_n)| < M(A)$$

for any finite set of non-overlapping subarcs $A_1, \ldots, A_n$ of A.

If the right swerve of A has bounded variation, then the left swerve has too, because a difference of $|s_r(A_i)|$ and $|s_1(A_i)|$ can originate only from the interior points of A_i whose complete angle is different from 2π. Therefore we simply say that the swerve of A has bounded variation. Strictly speaking we discussed swerve only on convex surfaces, but since we know that every manifold with non-negative curvature behaves locally like a convex surface, see (19.1), swerve and its variation are defined on such manifolds. With the previous notations we can now formulate the

(20.4) GENERAL GLUING THEOREM. *Let $M_1, \ldots, M_n$ be manifolds with non-negative curvature and let the swerve have bounded variation on any subarc of a C_j^i. The manifold M obtained by gluing the domains $\bar{D}_i \subset M_i$ together has non-negative curvature if, and only if, for any subarc A of two identified sections of C_r^i, C_s^j the sum of the swerves of A in M_i towards $\bar{D}_i$ and in M_j towards $\bar{D}_j$ is non-negative and for any point p belonging to more than two $\bar{D}_i$ the sum of the angles in these D_i at p is at most 2π.*

The angles exist because the C_j^i have definite directions towards both sides, see Section 15. The idea of the proof is the same as in the special theorem (19.4). However, a complete proof with all details is very long and, apparently, has never been published; for a sketch see A., pp. 362–364.

Here we discuss only one very interesting application of (20.4). In Section 15 we considered convex domains on convex surfaces. We generalize this concept slightly by admitting *relatively convex* domains. The closure F of an open set is called relatively convex if any two interior points of F can be connected by a shortest join relatively to F which lies entirely in the interior of F. If F is homeomorphic to a disk and relatively convex then its boundary B_F has the property that the swerve of any subarc of B_F towards F is non-negative. The proof is the same as that for (15.6). We saw in (11.9) that the closure of a convex cap, briefly a closed convex cap, is relatively convex. We can now show that this is, from the intrinsic point of view, the only type of relatively convex disk, provided we *widen the concept of cap* in the same way as in Section 16; namely, call cap K any non-complete convex surface homeomorphic to E^2 whose boundary is a closed plane convex curve B, such that the perpendicular projection of K on the plane of B lies inside or on B.

(20.5) *A relatively convex domain on a manifold with non-negative curvature which is homeomorphic to a disk is isometric to a closed convex cap (in the wider sense).*

This will evidently follow from

(20.6) THEOREM. *A domain F on a manifold with non-negative curvature which is homeomorphic to a disk is isometric to a closed convex cap (in the wider sense), if every subarc of its boundary B_F has non-negative swerve towards F.*

For a proof we observe first that the swerve of B_F towards F has bounded variation because it has always the same sign. It follows that B_F is rectifiable. [9]

Take a circular semi-cylinder C whose bounding circle B_C has the same length as B_F. Since the geodesic curvature of B_C on C is zero, we can apply (20.4): identifying B_F and B_C we obtain a manifold $M = F \cup C$ with non-negative curvature. M is homeomorphic to E^2 and finitely compact, hence can by (20.1) be realized as an open convex surface K in E^3. Denote by B, K_F and K_C the parts of K corresponding to $B_F = B_C$, F and C respectively.

K_C is also a semi-cylinder (of course, in general, not circular). For C being developable on a plane the same holds for K_C. A rectilinear generator L of K_C has the property that any subarc of it is a shortest connection. The image of L on C must have the same property, but the generators of C are the only curves with this property. Hence the generators of K_C and C correspond to each other, whence it easily follows that K_C is a cylinder. The curve B_C (in either orientation) is a shortest curve on C in its free homotopy class, hence the same holds for B on K_C. Therefore B must be a plane section of K_C perpendicular to the generators.

The supporting planes of K_C at interior points of K_C are also supporting planes of K. Thus the perpendicular projection of

[9] A simple proof is contained in Pogorelov [7], Chapter I, § 1.

$K_F - B$ on the plane of B lies inside or on B and $K_F - B$ is a convex cap in the wider sense.

We could easily have avoided using (20.4), which we did not prove, by approximating F with convex polygonal domains bounded by polygons inscribed to B_F, and applying the special theorem (19.4) to these, see A., pp. 368, 369.

An interesting corollary of (20.6) is the following answer to a problem raised by E. G. Straus:

(20.7) *A simple closed geodesic G on a closed convex surface C can be shrunk continuously on C to a point p through curves G_t, $0 \leqq t \leqq 1$, $G_0 = G$, $G_1 = p$, such that $\lambda(G_t)$ is a non-increasing function of t.*

For, either of the two domains on C bounded by G_0 satisfies the hypothesis of (19.9), hence is isometric to a cap K, whose boundary B corresponds to G_0. The plane sections of K parallel to the plane of G_0 provide the desired family G_t, with an obvious modification if the supporting plane of K parallel to the plane of B has more than one point in common with K.

21. Monotypy of closed surfaces

The monotypy of closed convex surfaces is one of the classical problems of differential geometry, hence a very brief outline of its history will be of interest.

The monotypy of the sphere among sufficiently smooth surfaces of positive curvature was conjuctured as early as 1838 by Minding, but was proved only in 1899 by Liebmann [1] and by Minkowski whose methods yield, as we saw, the general monotypy of the sphere, i.e., among all convex surfaces. In 1927 Cohn-Vossen [1] proved monotypy for closed surfaces within the class of piecewise analytic surfaces with positive curvature. In 1934 he weakened piecewise analyticity to class C^3. For the latter result Herglotz [1] found in 1942 his widely known proof, which is so simple that it can be found in text books and which contributed an essential idea to the general monotypy theorem. The smoothness hypotheses were further weakened by others. But, except for the already mentioned result of Olovyanishnikov [1], who proved general monotypy for polyhedra, all these results left *the disconcerting possibility that, for example, an ellipsoid might not be monotypic.* By theorem (5.2) any convex surface intrinsically isometric to an ellipsoid is at least of class C^1, so that this eventuallity could not be regarded as pathological.

The decisive new step was done by Pogorelov [1] (1949, a sketch appeared in 1948), who proved monotypy for closed surfaces with bounded specific curvature (see Section 4). In contrast to the previous hypotheses this condition is *intrinsic, so that monotypy in this class of surfaces entails general monotypy.*[10] Under the hypothesis of bounded specific curvature P. [1] also proves monotypy for open surfaces of curvature 2π (with a restriction which he later eliminated), for convex caps among convex caps, and for non-complete convex surfaces with total curvature 4π bounded by a finite number of, not necessarily plane, curves whose subarcs have, towards the surface, non-positive swerves.

Moreover, the basic idea of [1] enabled Pogorelov in 1951 to prove monotypy for general closed convex surfaces, and finally in [8], for general open convex surfaces with curvature 2π. [11] The very intricate and long proof for closed surfaces is now available in German as a little book, P. [7]. Here we merely outline briefly the main ideas, principally of P. [1], and give in detail only one particularly interesting step common to [1] and [7]. [12]

[10] Pogorelov proves in [3], see Section 23, that a, not necessarily complete, convex surface C intrinsically isometric to a smooth convex surface C_0 with positive curvature is itself smooth, although in general less so, but C is analytic with C_0. Hence analyticity, and to a certain degree smoothness, are in conjunction with positive curvature intrinsic properties. At present this result, together with Herglotz' method, cannot replace P. [1] for smooth surfaces, because P. [3] uses one of the main results of [1], namely the monotypy of caps among caps.

[11] That open surfaces with curvature less than 2π are never monotypic was proved earlier by Olovyanishnikov [2].

[12] Many theorems of P. [1] are, under somewhat stronger hypotheses and correspondingly simpler proofs, also found in P. [3].

(21.1) THEOREM. *Closed convex surfaces are monotypic.*

The following is to be proved: let $\varphi : p^1 \rightarrow p^2$ be a mapping of the closed convex surfaces C_1 on the closed convex surface C_2 which preserves the intrinsic distances:

(21.2) $$(p^1 q^1, C_1) = (p^1 \varphi q^1 \varphi, C_2) = (p^2 q^2, C_2).$$

Then

(21.3) $$|p^1 - q^1| = |p^1 \varphi - q^1 \varphi| = |p^2 - q^2|.$$

We may assume that φ preserves the orientation, for we may replace C_2 by its image under reflection in a plane. We assume further that C_1 and C_2 are parametrized in such a way that the same parameter values u, v on C_1 and C_2 yield points which correspond under φ.

Assume that C_1, and hence C_2, has bounded specific curvature. Then it follows from (4.9) that C_1 and C_2 cannot be doubly covered plane domains and from (5.2) that they are of class C^1, hence possess with suitable parameters u, v vector representations $\mathfrak{x}(u, v)$, $\mathfrak{y}(u, v)$ with $\mathfrak{x}_u \times \mathfrak{x}_v \neq 0$, $\mathfrak{y}_u \times \mathfrak{y}_v \neq 0$. Moreover, we conclude from (3.10), (4.8), and (4.9) the existence of a (u_0, v_0) such that $\mathfrak{x}(u_0, v_0)$ and $\mathfrak{y}(u_0, v_0)$ are Euler points with the same curvature $K_E > 0$. We place C_1 and C_2 into "canonical position," i.e., $\mathfrak{x}(u_0, v_0) = \mathfrak{y}(u_0, v_0)$, C_1 and C_2 have at this point the same tangent plane and the same interior normal; finally, directions on C_1 and C_2 at (u_0, v_0) corresponding under φ (which is conformal) coincide. We take the common tangent plane as (x_1, x_2)-plane and the common interior normal as positive x_3-axis.

In the general case, i.e., without the assumption of bounded specific curvature, (21.3) is very simple when both C_1 and C_2 are doubly covered domains, so that we may assume C_1 to be non-degenerate. One of the principal steps in both cases, and the only one which we will carry out in detail, consists in the reduction of the assertion to the case where the two surfaces are close together by showing:

(21.4) *If* C_1 *and* C_2 *satisfy* (21.2) *but not* (21.3) *then there exists either a* C'_1 *arbitrarily close to* C_1 *or a* C'_2 *arbitrarily close to* C_2

such that C_i *and* C'_i ($i = 1$ or 2) *satisfy* (21.2) *but not* (21.3).

If we apply this fact to C_1 and a degenerate C_2 then we obtain either a C_1' close to C_1 or a C'_2 close to C_2 satisfying (21.2) and not (21.3). In the latter case C'_2 is non-degenerate, and applying (21.4) once more to C_1 and C'_2 we obtain two non-degenerate surfaces which are close together and satisfy (21.2) but not (21.3). The substitute for "canonical position" in the general case used in P. [7] is involved and will not be discussed here.

In both cases the vector function

$$\mathfrak{w}(u, v) = \frac{\mathfrak{x}(u, v) + \mathfrak{y}(u, v)}{\mathfrak{x}^2(u, v) - \mathfrak{y}^2(u, v)}$$

plays a fundamental role. It makes sense only when $\mathfrak{x}^2(u, v) \neq \mathfrak{y}^2(u, v)$, but *the case* $\mathfrak{x}^2(u, v) \equiv \mathfrak{y}^2(u, v)$ *is simple*: Let p^1, q^1 be points of C_1 different from $z = (0, 0, 0) = \mathfrak{x}(u_0, v_0)$ and A_1 the arc not containing z in the plane section of C_1 through z, p^1, q^1. We form the cone

$$\bigcup_{y \in A_2} R(z, y), \quad A_2 = A_1 \varphi,$$

and unfold it on a plane. Then A_2 becomes a curve A'_2 which because of $\mathfrak{x}^2(u, v) \equiv \mathfrak{y}^2(u, v)$ is congruent to A_1. The distance of the end points p^2, q^2 of A_2 is at most as large as the distance of the end points of A'_2 which equals $|p^1 - q^1|$. Thus $|p^2 - q^2| \leqq |p^1 - q^1|$ and similarly $|p^1 - q^1| \leqq |p^2 - q^2|$.

We proceed under the assumption that C_1 and C_2 have bounded specific curvature. If $\mathfrak{w}_u \times \mathfrak{w}_v \neq 0$ for all u, v with $\mathfrak{x}^2(u, v) \neq \mathfrak{y}^2(u, v)$ then $\mathfrak{w}(u, v)$ represents, where defined, a surface F which for smooth $\mathfrak{x}(u, v)$, $\mathfrak{y}(u, v)$ is easily seen to have *non-positive curvature*. If u, v approach u', v' with $\mathfrak{x}^2(u', v') = \mathfrak{y}^2(u', v')$ then $|w_3(u, v)| \to \infty$. This is obvious for $(u', v') \neq (u_0, v_0)$, and follows for $(u', v') = (u_0, v_0)$ from the canonical position of C_1 and C_2 which guarantees that $\mathfrak{x}^2(u, v) - \mathfrak{y}^2(u, v)$ becomes small of order 3, whereas, because of $K_E > 0$, the numerator $x_3(u, v) + y_3(u, v)$ becomes small of order 2.

Consequently, if h is small, the region $-h < x_3 < h$ will not contain points of F. We may assume that F contains points in $x_3 > 0$ (otherwise we interchange C_1 and C_2) and draw a plane $x_3 = H > 0$ such that $0 < x_3 < H$ cuts a bounded piece F' off F. For sufficiently large $k > 0$ the paraboloid

$$P : z = x^2 + y^2 - k.$$

contains F' in its interior. We now deform F' affinely:

$$P_t : z = (1 - t)(x^2 + y^2) - (1 - t)k + tH, \quad 0 \leq t \leq 1.$$

Then P_t tends to $z = H$, and hence must for some t_0 touch F'. But at a point of contact F' would have positive curvature.

There are several *serious obstacles* in carrying this idea out. In the first place the hypothesis $\mathfrak{w}_u \times \mathfrak{w}_v \neq 0$ will not be satisfied, no matter how smooth C_1 and C_2 are, unless C_2 is close to C_1. This difficulty is overcome in both the special and the general cases by applying (21.4).

Secondly, F will, even when C_1 and C_2 have bounded specific curvature, not have a curvature everywhere. However, it *will behave like a surface with non-positive curvature for those u, v for which $\mathfrak{x}(u, v)$ and $\mathfrak{y}(u, v)$ are Euler points of C_1 and C_2.* By (3.8) the projection M on $x_3 = 0$ of the points of F which correspond to u, v for which not both $\mathfrak{x}(u, v)$ and $\mathfrak{y}(u, v)$ are Euler points has measure 0. Only if the paraboloid P_{t_0} touches F at a point whose projection on $x_3 = 0$ does not lie in M, does our argument apply. Therefore Pogorelov constructs instead of P a convex surface $x_3 = f(x_1, x_2)$ which is defined in the whole (x_1, x_2)-plane, contains F' in its interior and is such that, if subjected to a similar affine deformation as P, its *contact with F' cannot take place at a point over M.* The reasoning is similar to the last part in the proof of theorem (5.6). The construction of $f(x_1, x_2)$ is non-trivial; it is based on theorem (4.12).

We now come to (21.4), which is uninteresting as an assertion in as much as we are proving that its hypothesis is never satisfied. But the proof is very instructive as an application of Alexandrov's methods.

C_1 may be triangulated into convex triangles whose diameters are smaller than a given positive β and all whose angles are less than π. We first produce *a triangulation with further properties*: Let $\epsilon > 0$ be given. The complete additivity of excess or curvature entails that there is for each point p a $\delta(p) > 0$ such that $\nu(U(p, \delta(p)) - p) < \epsilon$. For a suitable finite number of points $p_1, \ldots, p_n$ the sets $U(p_i, \delta(p_i))$ cover C_1. There is a positive β' such that each triangle of diameter less than β' lies entirely in one of the $U(p_i, \delta(p_i))$. Then the excess of a triangle with diameter less than β' which does not contain a p_i in its interior is at most ϵ, and if a p_i lies in the interior, then segments from p_i to the vertices decompose the triangle into 3 triangles each of which has excess less than ϵ.

We can further reach that each segment which is a side of a triangle in the triangulation is the unique segment connecting its end points. For if Δ is a triangle in a triangulation satisfying the previous conditions, we draw the three medians. If these are concurrent then the 4 triangles obtained are adequate because of the inclusion property. Otherwise the medians decompose Δ into 4 triangles and 3 convex quadrangles. In each of the quadrangles we draw both diagonals and thus decompose Δ into 16 triangles with the desired property.

Consider a triangulation T_1 of C_1 which satisfies all these qualifications: *each Δ_1 in T_1 has diameter less than a given $\beta < 0$, $\epsilon(\Delta_1) < \epsilon$, all angles in Δ_1 are less than π, and each side of Δ_1 is the unique segment connecting its end points.* Under φ there corresponds to T_1 a triangulation T_2 of C_2 with the same properties. We take a finite set N_1 of points on C_1 which contains all vertices of T_1 and its image $N_2 = N_1\varphi$ on C_2. The convex closures of N_1 and N_2 are convex polyhedra P_1 and P_2. We connect the points of P_i corresponding to the vertices of T_i, $i = 1, 2$, by segments in the same way as for T_i. If N_i is sufficiently dense then the fact that sides of the triangles in T_i are unique segments connecting their end points entails that the triangulation T'_i of P_i is combinatorially the same as that of C_i, and that the curvatures of these triangles are small with ϵ. To each triangle Δ_i in T_i there corresponds a triangle Δ'_i in T'_i.

We deform P_i continuously as follows: if one of the triangles

Δ'_i of T'_i contains more than one corner (= proper vertex) of P_i in its interior, we take two vertices, say a_1 and a_2, connect them by a segment $T(a_1, a_2)$ on P_i, cut P_i open along $T(a_1, a_2)$ and insert two congruent triangles $a'_1 a'_2 x'$ and $a''_1 a''_2 x''$ exactly as in the proof of (17.4) and obtain a continuous transition of the polyhedral metric on P_i into a polyhedral metric in which the triangle corresponding to Δ'_i contains one corner less than Δ'_i. The curvature of Δ'_i does not change during this process. Each of these metrics is realizable as a polyhedron by theorem (17.1), and uniquely so by Cauchy's theorem. Hence the polyhedron will change continuously (if we make some agreement on its position as in Section 17).

Proceeding in this manner we deform P_i continuously into a polyhedron in which each triangle corresponding to a triangle in T_i or T'_i *contains at most one corner in its interior.* A triangle containing one corner is isometric to the lateral surface of a pyramid. If we connect the apex of the pyramid with an interior point of the base by a segment and let the apex slide along the segment we obtain, first abstractly, a continuous deformation of the polyhedral metric into a polyhedral metric with no corner inside the triangle. As above we can realize this process by a continuous deformation of an actual polyhedron in E^3.

Thus we obtain altogether *a continuous deformation of* P_i, $i = 1, 2$, *into a polyhedron* P_i^0 *such that each triangle* Δ_i^0 *on* P_i^0 *corresponding to a triangle* Δ'_i of T'_i *is isometric to a plane triangle.* It must be kept in mind that the Δ_i^0 will in general not be an an actual plane triangle; P_i^0 originates from these triangles by gluing.

All the polyhedra carry triangulations corresponding to T_1. We introduce a *deviation* for two such polyhedra Q', Q'' by

$$d(Q', Q'') = \inf_{\alpha} \max_{v'} |v' - v''\alpha|,$$

where α traverses all motions of E^3 including reflections, v' traverses the vertices of the triangulations of Q' corresponding to T_1 and $v''\alpha$ is the vertex corresponding to v' on $Q''\alpha$.

If the set N_1 is sufficiently dense, then the sides of a triangle

Δ_i in T_i differ little from the corresponding sides of Δ'_i in T'_i. Since Δ_1 and Δ_2 have equal sides, the sides of Δ'_1 and Δ'_2 differ little, so that for dense N_1 the deviation $d(P_1^0, P_2^0)$ is as small as desired.

The hypothesis that C_1 and C_2 do not satisfy (21.3) yields $d(P_1, P_2) \geqq M > 0$ as soon as the triangulation T_i is sufficiently fine, N_i sufficiently dense and ϵ sufficiently small. The deviation changes continuously under our deformation, hence for a given $\delta < M$ either P_1 or P_2 (or both) passes under the deformation, a position P_1^* or P_2^* such that

$$d(P_i^*, P_i) = \delta.$$

Finally we replace T_i, N_i, ϵ by a sequence of triangulations $T_{i,\nu}$, which get finer and finer, N_i by a sequence $N_{i,\nu}$ which becomes denser and ϵ by a sequence $\epsilon_\nu \to 0$. For each ν we have polyhedra $P^*_{1,\nu}$ and $P_{1,\nu}$ say, with $d(P^*_{1,\nu}, P_{1,\nu}) = \delta$, and $P_{1,\nu} \to C_1$ whereas $P_{1,\nu}$ tends (possibly after passing to a subsequence) to a surface C'_1 which satisfies our assertion because δ is a given positive number less than M and independent of ν.

22. Other monotypy theorems. Deformations

Theorem (21.1) leads easily to the following result on non-complete surfaces:

(22.1) THEOREM. *Let C be a non-complete convex surface with curvature $\nu(C) = 4\pi$ whose boundary consists of a finite number of disjoint curves $J_1, \ldots, J_m$ which are either closed Jordan curves or cuts, i.e., Jordan arcs traversed back and forth. If every subarc of a J_i has non-positive swerve towards C, then C is monotypic.*

We emphasize that we do *not* assume that the J_i be plane curves, which is essential for the corresponding theorem in the classical approach through differential equations. On the other hand it seems, according to a statement (without a reference) in A., p. 488 that no hypotheses on the boundary of C are necessary: any surface C with $\nu(C) = 4\pi$ is monotypic.

For a proof of (22.1) we consider a surface C' related to C

by an intrinsically isometric mapping $\varphi : p \to p'$, $(pq, C) = (p'q', C')$ and reduce (22.1) to (21.1) by showing that *φ can be extended to an intrinsically isometric mapping of the boundaries K, K' of the convex closures of C and C' respectively.*

$K - C$ consists of a finite number of disjoint closed sets $F_1, \ldots, F_m$ each bounded by (or in case of a cut coinciding with) a J_i. If J_i is a cut then the swerve of any subarc of J_i towards both sides vanishes because of $\nu(C) = 4\pi$ and (15.4). Hence J_i is a quasigeodesic with vanishing swerves, see Section 15. If J_i is a closed Jordan curve then F_i is by (20.6) isometric to a closed cap, see (20.6) and because $\nu(C) = 4\pi$ implies $\nu(F_i) = 0$, the cap is a *plane convex domain.* Similarly we have on K' corresponding to $F_1, \ldots, F_m$ sets $F'_1, \ldots, F'_m$ bounded by curves $J'_1, \ldots, J'_m$ and either J'_i is a quasigeodesic or F'_i is isometric to a plane convex domain.

If J_i is a quasigeodesic then J'_i is too, because otherwise J'_i would be intrinsically a plane convex curve with the curvature concentrated at the two points corresponding to the end points of J_i.

If J_i and J'_i are not quasigeodesics then the plane domains, to which F_i and F'_i are intrinsically isometric, are congruent, because swerve is an intrinsic concept and preserved under φ. The swerve of a subarc of a plane convex curve as a function of the arc length determines the curve up to motions.

It is now clear that φ can be extended to an isometry of K on K'. Since theorem (21.1) is so much harder to prove for the general case than for bounded specific curvature, we observe that K has, trivially, bounded specific curvature if C does: For any Borel set M on $K - C$ we have $\nu(M) = 0$ and if $A(C \cap M) \neq 0$ then

$$\frac{\nu(M)}{A(M)} = \frac{\nu(M \cap C)}{A(M)} \leqq \frac{\nu(M \cap C)}{A(M \cap C)}.$$

The assumption $\nu(C) = 4\pi$ in (21.5) is essential. The strongest known statement in this direction is due to Leibin [1]:

(22.2) THEOREM. *Let C be a convex surface bounded by a finite* *number of disjoint closed Jordan curves. If the largest convex set whose boundary contains C is distinct from the smallest* (i.e., the convex closure of C) *then C is deformable.*

Leibin constructs the deformation by means of the gluing theorem (20.4), but does not assert continuity of the deformation. This follows, however, from the — then not known — general monotypy theorem (21.1). Theorem (22.2) implies in particular that C is deformable if it originates from a closed convex surface by removing the closure $\bar{D}$ of an open set D with $\nu(D) > 0$. The proof of (22.2) is long and we give here merely the *very special case of the last statement where $\bar{D}$ is a proper geodesic triangle abc.*

Because $\epsilon(D) = \nu(D) > 0$ at least one of the angles α, β, γ in $\bar{D}$ is greater than the corresponding angle in a euclidean triangle with the same sides. The convexity condition shows easily that actually all three are greater. In E^2 we form a convex hexagon $a'd'b'e'c'f'$ such that

$$(23.3)\qquad \begin{gathered} |a'-d'| + |d'-b'| = ab, \quad |b'-e'| + |e'-c'| = bc, \\ |c'-f'| + |f'-a'| = ca \\ \angle f'a'd' \leqq \alpha, \quad \angle d'b'e' \leqq \beta, \quad e'c'f' \leqq \gamma. \end{gathered}$$

These conditions guarantee that the hexagon can be glued to $\bar{C}$ such that a' falls on a, b' on b and c' on c and that the resulting manifold has non-negative curvature, see (19.4). Since the manifold is homeomorphic to S^2 it can by (18.4) be realized as a convex surface K. The part C' of K corresponding to C is intrinsically isometric, but not congruent, to C because of the angles occurring at the points corresponding to d', e', f'. These points as well as the angles $\angle f'a'd'$, . . . in (23.3) can be varied continuously within certain limits. The monotypy theorem (21.1) implies that C' can be placed so that it varies continuously with these points and angles.

Theorem (22.2) implies that *caps are never rigid,* but this is very easily seen directly from the gluing theorem. Let K be a convex cap with boundary B. We imbed B in any continuous family B_t, $0 \leqq t \leqq 1$, $B_0 = B$, of plane convex curves with the same length as B. We glue the plane domain $\bar{D}_t$ bounded by B_t

to $\bar{K}$. Since the swerve of any subarc of B towards K and of any subarc of B_t towards $\bar{D}_t$ is non-negative, see Section 20, the resulting manifold has by (20.4) non-negative curvature and is by (18.4) monotypic. The part K_t corresponding to K on a realization of $\bar{K} \cup \bar{D}_t$ is intrinsically isometric, but in general not congruent, to K. With a proper normalization of their positions the K_t represent a deformation of K. Thus we can only expect:

(22.4) THEOREM. *Convex caps are monotypic among convex caps.*

The difficulties of this theorem are similar to those of (21.1), however, large parts of the proof of (21.1) apply also to (22.4). Here we give merely a brief outline for the case of bounded specific curvature. A simple proof under smoothness hypotheses is found in Grotemeyer [1].

Let K and K' be two equally oriented convex caps related by an intrinsically isometric mapping $\varphi : p \rightarrow p'$. We place K and K' so that their boundaries lie in a plane $x_3 = h > 0$ and the surfaces themselves in $0 < x_3 < h$. Choose parameters u, v on K and K' such that points corresponding under φ have the same parameters. If u, v are suitably chosen then the vector representations $\mathfrak{x}(u, v)$ and $\mathfrak{x}'(u, v)$ of K and K' are of class C', see (5.3).

Denote by K'' the image of K' under reflection in $x_3 = 0$, so that $\mathfrak{x}''(u, v) = (x'_1(u, v), x'_2(u, v), -x'_3(u, v))$ represents K'', and form

$$\mathfrak{w}(u, v) = \mathfrak{x}(u, v) + \mathfrak{x}''(u, v).$$

If $\mathfrak{w}_u \times \mathfrak{w}_v \neq 0$ this is a surface of non-positive curvature, but since its boundary lies in $x_3 = 0$, the surface lies in $x_3 = 0$, or $x_3(u, v) \equiv - x''_3(u, v) \equiv x'_3(u, v)$.

The rest is simple: two given points p, q of K lie in a section of K by a plane normal to the (x_1, x_2)-plane. φ maps the arc from p to q of this section on an arc A' on K' from p' to q'. We unfold the cylinder with generators parallel to the x_3-axis through A' on a plane and obtain from A', because of $x_3(u, v) \equiv x'_3(u, v)$ a plane curve $\bar{A}$ congruent to A. The end points of $\bar{A}$ have therefore distance $|p - q|$ and this at least as large as $|p' - q'|$, the distance of the end points of A'. Similarly $|p' - q'| \geqq |p - q|$.

The last proof yields clearly a slightly more general theorem than (22.4), see P [1].

(22.5) *Let the convex surfaces K and K' be both represented in the form $x_3 = f(x_1, x_3)$ and have boundaries B, B' which are closed Jordan curves. If there is an intrinsically isometric mapping of K on K' which induces a mapping of B on B' for which corresponding points have equal x_3-coordinates, then K and K' are congruent.*

We will not discuss the proof of the monotypy theorem for open surfaces at all and mention only that it uses Herglotz' idea explicitly, see P. [8].

(22.6) THEOREM. *Open convex surfaces with curvature 2π are monotypic.*

Open surfaces with curvature less than 2π are not monotypic, see below, but we may expect from theorem (4.11) that they become monotypic, if we keep their limit cones fixed. The most complete theorem is obtained by introducing the term "*geodesic ray*" for a curve $p(s)$, $0 \leqq s < \infty$ as an open convex surface C for which (with the notation of Section 10)

$$\lambda_0^s(p) = s = p(0)p(s),$$

in words, any subarc of $p(s)$ is a segment. $p(s)$ as curve $p(s)$ in E^3 possesses everywhere a right-hand tangent

$$p'_r(s) = \lim_{h\to 0+} [p(s+h) - p(s)]\, h^{-1}, \quad |p'_r(s)| = 1,$$

see theorem (12.1). Olovyanishnikov [2] showed that $p'_r(s)$, laid off from the apex a of a limit cone D of C tends to a generator $R(p)$ of D. Pogorelov proves, see [1] and [7]:

(22.7) THEOREM. *Let C and C' be equally oriented surfaces with curvature less than 2π and φ an intrinsically isometric mapping of C on C' which preserves the orientation. If C and C' have the same non-degenerate limit cone D and if there are geodesic rays $p(s) \in C$ and $p'(s) \in C'$ which correspond under φ such that $R(p) = R(p')$, then C' originates from C by a translation.*

Olovyanishnikov [2] showed:

(22.8) THEOREM. *Let C be an open convex surface with cur-*

vature less than 2π, a non-degenerate limit cone D, and $p(s)$ a geodesic ray on C. If D' is a non-degenerate convex cone with the same curvature as D' and R' is any generator of D', then a convex surface C' intrinsically isometric to, and with the same orientation as, C exists which has D' as limit cone and $R' = R(p')$ where $p'(s)$ corresponds to $p(s)$ under the isometry.

By (22.7) C' is unique and we conclude that C' will change continuously with D' and R'. Theorems (22.6) and (22.7) remain true for surfaces obtained from a complete open surface through cutting out holes of the type occurring in (22.1).

23. Smoothness of realizations

In Sections 19 and 20 we established the existence of realizations of given metrics with non-negative curvature. The deformation processes of the last section provide for many types of metrics a great variety of realizations. Now we raise the question whether the realization is necessarily smooth when the given metric is smooth. The only facts in this direction, which we obtained, so far, are the implications of theorems (5.1) and (5.4), in particular:

(23.1) *A convex surface whose specific curvature is bounded away from 0 and ∞ is strictly convex and of class C^1.*

In comparison with the theory of Sections 18–22 the results on smoothness which go beyond (23.1) are, in spite of considerable efforts of very good mathematicians, rather incomplete, because just the geometrically most interesting cases have eluded us.

There are few significant geometric formulae which contain higher derivatives of the coefficients E, F, G of the line element than the first, and none that contain (unavoidably) higher derivatives than the second. An example of a surface $\mathfrak{x}(u, v)$ with E, F, G of class C^3 and positive curvature where $\mathfrak{x}(u, v)$ is of lower class is not known. Should such a surface exist (which is not likely[13] then the case where E, F, G are of class C^3 would be

[13] The author does not even know of an example of a convex surface where the E, F, G are of class C^m, $m \geqq 2$, $EG - F^2 > 0$, the curvature is positive and $\mathfrak{x}(u, v)$ is not of class C^{m+1}.

of geometric interest. But our present information begins only with class C^4.

The situation is much more satisfactory from the analyst's point of view. For we will see that $x(u, v)$ is of class C^m, whenever the E, F, G are of class C^m, $m \geqq 4$, $EG - F^2 > 0$, and the curvature is positive.

Since the plane can be "folded" along a line, a convex surface which contains pieces of vanishing curvature will in general not be of class C^1 no matter how smooth the E, F, G are. It is possible that smooth metrics with non-negative curvature always possess smooth realizations, but nothing is known in this respect (beyond (5.2)).

For a rigorous formulation of the results we repeat some more or less standard definitions. For the definition of the class $C^{m+\alpha}$, m an integer, $0 < \alpha < 1$, we refer the reader to Section 5. It is here convenient to admit $\alpha = 0$, identifying C^{m+0} with C^m. A *coordinate domain* K^U on a convex surface K is obtained by mapping an open set on K topologically on an open set U of the (u, v)-plane and assigning to a point of K^U as coordinates u, v those of its image in U.

We say that K^U *is of class* $C^{m+\alpha}$, $m \geqq 1$, *or analytic*, if the vector representation $\mathfrak{x}(u, v)$ of K^U as surface in E^3 in terms of the $(u, v) \in U$ has components which are of class $C^{m+\alpha}$ or analytic. If $\mathfrak{x}_u \times \mathfrak{x}_v \neq 0$ then K^U is *regular*. *The intrinsic metric of* K^U *is of class* $C^{m+\alpha}$, $m \geqq 0$, *or analytic or regular* if $ds^2 = E(u, v)du^2 + 2F(u, v)dudv + G(u, v)dv^2$ and the E, F, G are of class $C^{m+\alpha}$, or analytic, or $EG - F^2 > 0$, respectively. This means that, necessarily unique, functions E, F, G with the respective properties exist such that for any curve $(u(t), v(t))$, $\alpha \leqq t \leqq \beta$, in U of class C^1 the length of the corresponding arc A on K^U is given by

$$\lambda(A) = \int_\alpha^\beta (Eu'^2 + 2Fu'v' + Gv'^2)^{1/2} dt.$$

If E, F, G are of class C^2 then the Gauss curvature can be evaluated by Gauss' formula. That this curvature be positive is then equivalent to saying that the specific curvature in a neighborhood of a given point of U be bounded away from 0. Although this formulation seems preferable in the framework of the present theory, we avoid it in order to conform with the familiar terminology.

The principal result is:

(23.2) THEOREM. *If for a coordinate domain K^U on a convex surface the intrinsic metric of K^U is regular and of class $C^{m+\alpha}$, $m \geqq 4$, $0 \leqq \alpha < 1$, (analytic) and has positive curvature, then K^U is regular,*[14] *and at least of class $C^{m+\alpha}$ (analytic).*

This result implies corresponding results in the large. Pogorelov says that the (intrinsic metric of the) convex surface K is of class $C^{m+\alpha}$, analytic, or regular if every point of K lies in a suitable coordinate neighborhood K^U which (whose intrinsic metric) is of class $C^{m+\alpha}$, analytic or regular. Usually a transition or consistency condition is added: in the overlap $K^U \cap K^{U'}$ of two coordinate domains each point has two pairs of coordinates u, v and u', v' and in $K^U \cap K^{U'}$ the u, v are continuous functions $u(u', v')$, $v(u', v')$ with a continuous inverse $u'(u, v)$, $v'(u, v)$. The transition condition requires for class $C^{m+\alpha}$ or analyticity that the coordinate domains can be chosen such that the four functions are of class $C^{m+\alpha}$ or analytic in each non-empty overlap $K^U \cap K^{U'}$. At any rate (23.2) implies for Pogorelov's and hence all the more for the usual definition:

(23.3) THEOREM. *If the intrinsic metric of a convex surface K is of class $C^{m+\alpha}$, $m \geqq 4$, $0 \leqq \alpha < 1$ (analytic) and regular and has positive curvature, then K is regular and at least of class $C^{m+\alpha}$ (analytic).*

This theorem under the assumption that the intrinsic metric be of class C^m, $m \geqq 5$ and the assertion that the surface is of class C^{m-1}, is due to Pogorelov [3]. The improvement on the hypothesis and the assertion of Pogorelov is obtained by the methods of Nirenberg [2], which we used for a similar purpose in theorem (5.6). The first important step toward (23.3) was the theorem

[14] Because $EG - F^2 = (\mathfrak{x}_u \times \mathfrak{x}_v)^2$ the regularity of K^U is equivalent to that of its intrinsic metric.

of Weyl-Lewy that an analytic line element on S^2 with positive curvature is realizable as an analytic convex surface. The monotypy theorem for closed convex surfaces (with bounded specific curvature) yields uniqueness and hence (23.3) for closed convex surfaces with analytic metrics. The result of Weyl-Lewy is not only of historical importance. Their method is essential also for (23.3). For Pogorelov does not discuss the given surface directly (as in the proof of (23.1)), but proves the existence of a smooth cap H_0 with the same intrinsic metric as a small given cap H on the given surface and then uses the monotypy of caps to assert that H is congruent to H_0 and hence smooth. Moreover, the existence proof for H_0 draws essential ideas from the Weyl-Lewy method. All these authors use in turn very essentially the ideas of S. Bernstein in his fundamental contributions (1904–1910) to the theory of elliptic partial differential equations.

The proof of (23.3) in P. [3] occupies two chapters (IV, V pp. 83–121) and the two appendices, (pp. 160–182). We must therefore restrict ourselves to a description of the basic ideas. One of the main steps is the proof of:

(23.4) THEOREM. *In a neighborhood U of the disk D: $u^2 + v^2 \leqq 1$ let an analytic line element with positive curvature be given for which the boundary B of D has everywhere positive geodesic curvature towards D. Let any geodesic issuing from $q = (0, 0)$ intersect B in a point whose distance from q measured along the geodesic is less than $K_M^{-1/2}$, where K_M is the maximum of the Gauss curvature in D. Then there exists an analytic convex cap which realizes D with the given line element.*

The assumption on the geodesic curvature of B is, of course, suggested by theorem (20.6). For a proof of (23.4) we introduce geodesic polar coordinates ϱ, β with center q, so that line element and curvature take the form

$$ds^2 = d\varrho^2 + c^2(\varrho, \beta)d\beta^2, \ K = -c^{-1}c_{\rho\rho}.$$

Since c^2 satisfies on a geodesic issuing from q, i.e., for fixed β, the differential equation

$$y''(\varrho) + Ky(\varrho) = 0, \quad y(0) = 0, \quad y'(0) = 1,$$

familiar arguments show that $c(\varrho, \beta)$ is for fixed β concave and increases in the interval $\pi/2K_M^{-1/2}$. Hence the polar coordinates are defined until we reach the limits of U or $\varrho = K_M^{-1/2}$ at any rate in a neighborhood of D.

Let $\varrho = \varrho(\beta)$ be the equation of β. It follows rather easily from the Gauss-Bonnet theorem that the geodesic curvature of the curve $\beta_\alpha : \varrho = \alpha\varrho(\beta)$, $0 < \alpha < 1$ at $(\alpha\varrho(\beta), \beta)$ is greater than that of B at $(\varrho(\beta), \beta)$, P [3], pp. **113**, **114**.

We map the closed subdomain D_α bounded by B_α on D by mapping the point ϱ, β in D_α on the point $\alpha^{-1}\varrho$, β in D. Then B_α goes into B and the mapping can be extended to a neighborhood of D_α on a neighborhood of D. The given metric ds^2 considered in a neighbornood of D_α becomes then a metric $ds_\alpha{}^2$ in a neighborhood of D and $ds_\alpha{}^2$ depends analytically on α; moreover we saw that the geodesic curvature of B for the metric $ds_\alpha{}^2$ is positive towards D.

(23.4) is based on a lemma, whose formulation and proof are made concise by saying briefly that an analytic line element with positive curvature defined in a neighborhood of D can be *realized as a cap beyond* D if a certain neighborhood of D can be realized as an analytic convex surface of which the part corresponding to D is a closed cap.

(23.5) *Lemma: In a neighborhood of D let an analytic line element*

$$ds_\alpha{}^2 = E_\alpha du^2 + 2F_\alpha du\,dv + G_\alpha dv^2, \ \alpha_0 \leqq \alpha \leqq \alpha_1,$$

be given which depends analytically on α, and such that B has for each α at each point positive geodesic curvature towards D. If $ds_{\alpha_0}{}^2$ can be realized as a cap beyond D, then $ds_\alpha{}^2$ can be so realized for all α.

Because of this lemma it suffices in our case to show that for some $\alpha \neq 0$ the metric $ds_\alpha{}^2$ is realiable as a cap beyond D, so that we reduce the problem to studying ds^2 in D_α, $0 < \alpha < 1$, for small α. If K_q is the curvature of ds^2 at q, denote by

$$d\tilde{s}^2 = d\varrho^2 + \tilde{c}^2(\varrho, \beta)d\beta^2$$

the line element in geodesic polar coordinates of a sphere S with radius $K_q^{-1/2}$. We consider in D the line element

$$d\tilde{s}_t^2 = d\varrho^2 + [tc + (1-t)\tilde{c}]^2 d\beta^2, \qquad 0 \leqq t \leqq 1.$$

For small $\alpha_0 > 0$ this metric has positive curvature in D_{α_0} and B_{α_0} has positive geodesic curvature for each $d\tilde{s}_t^2$. By (23.5) the metric ds^2 can be realized as a cap beyond D_{α_0} if ds_t^2 can for some t.

Denote by $\tilde{D}$ the domain of S corresponding to the domain D_{α_0} in the (ϱ, β)-plane. The projection B' of the boundary $\tilde{B}$ of $\tilde{D}$ from the center of S on the tangent plane P of S at the point corresponding to $\varrho = 0$ or q is a convex curve with positive curvature in P. In P we introduce polar coordinates r, δ such that $\delta = 0$ corresponds to $\beta = 0$. Let $r = f(\delta)$ be the equation of B'. Then the curve B'_s in P given by

$$r = f(\delta)[(1-s) + sf(\delta)]^{-1}, \qquad 0 \leqq s \leqq 1$$

has everywhere positive curvature and is for $s = 0$ the curve B' and for $s = 1$ the circle $r = 1$.

Let $D_{\alpha_0, s}$ be the convex domain on S obtained by projecting the convex domain in P bounded by B'_s on S (from the center of S). By the lemma $\tilde{D}$ with the metric of S is realizable as cap beyond $\tilde{B}$ if this is so for some $D_{\alpha_0, s}$ with the spherical metric. But this is obvious for $s = 1$ because we then have a spherical cap.

The *proof of the lemma* (23.5) is based on the theory of elliptic partial differential equations. The equation is *Darboux's equation* which is derived as follows:

Assume $\mathfrak{x}(u, v)$ to represent a surface with positive curvature and that u, v are geodesic parallel or polar coordinates, so that the line element has the form

$$ds^2 = du^2 + c^2(u, v)dv^2 = dx_1^2 + dx_2^2 + dx_3^2.$$

If we write

$$dx_1^2 + dx_2^2 = ds^2 - dx_3^2$$

and express that the left side has, as line element of a plane, curvature 0, we obtain, with the usual abbreviations

$$\frac{\partial x_3}{\partial u} = p, \quad \frac{\partial x_3}{\partial v} = q, \quad \frac{\partial^2 x_3}{\partial u^2} = r, \quad \frac{\partial^2 x_3}{\partial u \partial v} = s, \quad \frac{\partial^2 x_3}{\partial v^2} = t.$$

Darboux's equation [15]

$$\varphi(u, v, p, q, r, s, t) = r(t + A) - (s - B)^2 + C = 0$$

where

$$A = cc_u p - c^{-1} c_v q, \; B = c^{-1} c_u q, \; C = c^{-1} c_{uu} (c^2 - c^2 p^2 - q^2).$$

The discriminant of this equation is

$$\varphi_r \varphi_t - 4^{-1} \varphi_s^{\,2} = (t + A)r - (s - B)^2 = -C = -c^{-1} c_{uu} (c^2 - c^2 p^2 - q^2).$$

Now $-c^{-1} c_{uu}$ is the Gauss curvature and the second factor on the right is the discriminant of the form

$$dx_1^{\,2} + dx_2^{\,2} = ds^2 - dx_3^{\,2} = du^2 + c^2 dv^2 - (p\,du + q\,dv)^2.$$

Assume now that $\mathfrak{x}(u, v)$ is a convex cap H with positive curvature so placed that it lies in $x_3 > 0$ and its boundary lies in $x_3 = 0$. Then the ratio of the discriminants of the forms $dx^2 + dy^2$ and ds^2 equals θ, where θ is the angle between the tangent plane of H and $x_3 = 0$.

Therefore the discriminant $\varphi_r \varphi_t - 4^{-1} \varphi_s^{\,2}$ will be positive and bounded away from 0, if $K > K_0 > 0$ and $0 \leqq \theta < \pi/2 - \theta_0$, $\theta_0 > 0$. *Darboux's Equation is then elliptic.* We observe for a later application that the latter condition suffices, i.e., the boundary of H need not lie in $x_3 = 0$ or be parallel to it, as long as θ is bounded away from $\pi/2$.

The first step in the proof of (2.5) is justified by a general theorem of S. Bernstein [1] on elliptic partial differential equations: if $ds_{\alpha'}^{\,2}$, is realizable as a cap beyond D for some α' then the same holds for all α close to α', i.e., the set of α, for which $ds_\alpha^{\,2}$ is realizable as a cap beyond D, is open in $[\alpha_0, \alpha_1]$.

The second step consists in showing that $ds_\alpha^{\,2}$ is realizable if $ds_{\alpha_\nu}^{\,2}$ is realizable as a cap K_ν beyond D for a sequence of α_ν which tend to α. We place the caps K_ν such that they lie in $x_3 > 0$,

[15] See Darboux [1], vol. III, Livre VII, Chapitre IV, where the corresponding equation is derived for a general line element.

their boundaries lie in $x_3 = 0$ and contain the origin in the interior. If δ^2 is an upper bound for E_α, $|F_\alpha|$, G_α then

$$|\mathfrak{x}_\alpha(u'', v'') - \mathfrak{x}_\alpha(u', v')| \leq \delta[(u'' - u')^2 + (v'' - v')^2]^{1/2},$$

therefore the K_ν are uniformly bounded and equicontinuous and hence may be assumed to tend to a cap K of which we want to prove that it realizes $ds_\alpha{}^2$ and can be analytically continued.

Since the work of S. Bernstein the *method of a priori bounds* has been used for this purpose. Uniform bounds for the first derivatives of $\mathfrak{x}^\nu(u, v) = \mathfrak{x}_{\alpha_\nu}(u, v)$ are easily obtained from the bounds for E_α, $|F_\alpha|$, G_α. Bounds for the second derivatives offer here, as in all similar cases, the greatest difficulties; because of the intricacy of the argument we merely refer the reader to P. [3], pp. 88–99. There result *bounds for the absolute values of the second derivatives of the* $\mathfrak{x}^\nu$ which depend on the greatest lower bounds of the Gauss curvature of the K_ν, the geodesic curvatures of B as curve on K_ν, and of $E_{\alpha_\nu}G_{\alpha_\nu} - F_{\alpha_\nu}{}^2$ and also depend on the least upper bounds of the absolute values of E_{α_ν}, $|F_{\alpha_\nu}|$, G_{α_ν} and their derivatives up to order 4, and on the derivatives up to order 4 of the functions $u_\nu(s)$, $v_\nu(s)$ representing B in terms of arc length with respect to $ds_{\alpha_\nu}{}^2$. In terms of these same quantities there is also a bound $\pi/2 - \theta_0$, $\theta_0 > 0$ for the angles of the tangent plane of K_ν with $x_3 = 0$.

Using S. Bernstein's ideas Pogorelov [3], Appendix II obtains bounds for the higher derivatives of x in terms of the second derivatives and certain other quantities. We do not state his result, although it would suffice for the present analytic case, but rather the stronger results on which Nirenberg's method is based.

(23.6) *Let $z(u, v)$ be a solution of class C^2 of the non-linear elliptic partial differential equation*

$$\varphi(u, v, z, z_u, z_v, z_{uu}, z_{uv}, z_{vv}) = 0$$

in a certain open (u, v)-domain U. If φ possesses partial derivatives up to order m, and the mth derivatives satisfy Hölder conditions *with an exponent* β, $0 < \beta < 1$, *then* z *has derivatives up to order* $m + 2$ *and its* $(m + 2)$*nd derivatives satisfy Hölder conditions with the exponent* β *in every closed subdomain* D *of* U. *If* φ *is analytic then* z *is analytic.*

This theorem, under the additional hypothesis that the second derivatives of z satisfy Hölder conditions, is due to E. Hopf [1], the removal of this condition to Nirenberg [1], which also contains, under the hypothesis of (23.6) the theorem:

(23.7) *The second derivatives of* z *satisfy in* D *a Hölder condition whose coefficient and exponent depend only on*: 1) *the distance of* D *from the boundary of* U, 2) *the least upper bounds of the absolute values of the first and second derivatives of* z *in* U, 3) *the least upper bounds in* U *of the absolute values of the eight first derivatives of* F (*after* z *has been substituted into the formally calculated derivatives*), 4) *the least upper bounds in* U *of* $|\varphi_{z_{uu}}|^{-1}$, $|\varphi_{z_{vv}}|^{-1}$ *and* $(\varphi_{z_{uu}}\varphi^2_{z_{vv}} - 4^{-1}\varphi_{z_{uv}})^{-1}$.

These results are applied to the case at hand as follows: we replace the x_1- and x_2-axes by x'_1- and x'_2-axes which are close to, but not in one plane with, the x_3-axis. Then the angle of the tangent plane of K_ν with any of the new coordinate planes is bounded away from $\pi/2$. The covariant components $\bar{x}_1(u, v)$, $\bar{x}_2(u, v)$, $\bar{x}_3(u, v) = x_3(u, v)$ of $\mathfrak{x}(u, v)$ with respect to the new axes then satisfy, according to a previous remark, Darboux equations which are of the elliptic type. We apply (23.8) and Pogorelov's bounds on the second derivatives of the covariant components x_i^ν of x^ν and the angles which the tangent plane of K_ν forms with the new coordinate planes to obtain uniform Hölder exponents and constants for the second derivatives of the x_i^ν. Therefore $x_i(u, v) = \lim_\nu x_i^\nu(u, v)$, is of class C^2 and solves Darboux's equation. By (23.7) the $x_i(u, v)$ are analytic, hence K is analytic and can by a result of S. Bernstein be analytically continued beyond B.

We now outline the *proof of theorem* (23.2). Let $q = x(u_0, v_0)$ be an arbitrary point of K^U. By (23.1) a plane parallel and sufficiently close to the tangent plane of K^U at q will cut a cap H

off K^U whose intrinsic diameter is less than K_M^{-1}, where K_M is the maximum of the curvature on H. Let L_H be the boundary of H.

By inscribing in L_H a suitable geodesic polygon and rounding its corners off with small geodesic circles we obtain a curve L'' arbitrarily close to L_H. By the same process as in the proof of (23.4), (where we replaced $\varrho(\beta)$ by $\alpha\varrho(\beta)$, $0 < \alpha < 1$) we obtain from L'' a curve L' arbitrarily close to L_H which has at every point positive right and left geodesic curvatures towards q.

Let the metric of K^U be given by

$$ds^2 = E\,du^2 + 2F\,dudv + G\,dv^2.$$

We approximate E, F, G and their derivatives up to order 4 (inclusive), uniformly in any closed subset of U, by analytic functions E_n, F_n, G_n and their corresponding derivatives and form the line element

$$ds_n{}^2 = E_n du^2 + 2F_n dudv + G_n dv^2. \tag{23.8}$$

For large n the curvature $K_n(u, v)$ of $ds_n{}^2$ is arbitrarily close to the curvature $K(u, v)$ of ds^2 and L', now as curve in the (u, v)-domain, will have everywhere positive right and left geodesic curvatures towards (u_0, v_0) for the metric $ds_n{}^2$. Moreover the intrinsic diameter of the domain bounded by L' will be less than $K_{nM}^{-1/2}$, where $K_{nM} = \sup K_n(u, v)$, $(u, v) \in U_H$ and U_H is the subdomain of U corresponding to H.

If $\varrho = f_1(\beta)$ is the equation of L' in geodesic polar coordinates on K^U with center q, then a familiar argument from real variables yields an analytic function $f(\beta)$ arbitrarily close to $f_1(\beta)$ such that $\varrho = f(\beta)$ has everywhere positive geodesic curvature for ds^2 and hence for $ds^2{}_n$ for large n. The (ϱ, β)-domain bounded by L can be mapped conformally on the domain $D : x^2 + y^2 < 1$ such that $\varrho = 0$ goes into $x = y = 0$ and this mapping can be analytically continued beyond L.

In a sufficiently small neighborhood of $\bar{D}$ we assign to a point as coordinates ϱ, β those of its preimage. We map a small neighborhood of the domain on K^U bounded by the curve corresponding to L on a neighborhood of $\bar{D}$ by associating points with the same ϱ, β. Then the metric ds_n^2 given in H can be considered as a metric in a neighborhood of $\bar{D}$. By theorem (23.4) there is an analytic convex cap which realizes D with ds_n^2 as line element.

We now take a sequence L_j of curves of the same type as L which tend to L_H. Then we have for large n and all j an analytic cap $H_{n,j}$ realizing ds_n^2 inside L_j as an analytic cap. On $H_{n,j}$ we introduce coordinates u, v by assigning to the point ϱ, β in $H_{n,j}$ the same u, v as to the point ϱ, β on H. Then (23.8) is the line element of $H_{n,j}$. We place the $H_{n,j}$ again in $x_3 > 0$ with their boundaries in $x_3 = 0$ and surrounding the origin. Then a suitable subsequence $H_k = H_{n_k, j_k}$, $n_k \to \infty$, $j_k \to \infty$, converges to a cap which is by (11.3) intrinsically isometric, and hence by (22.4) congruent, to H.

Next we take a subdomain H' of H which contains q in its interior and whose closure lies in H. To H' there corresponds (in terms of u, v) a subdomain H'_k of H_k whose distance from $x_3 = 0$ is bounded away from 0. Consequently the angles of the tangent planes of H'_k will be uniformly bounded away from $\pi/2$.

As before we replace the x_1- and x_2-axes by x'_1 and x'_2-axes close to the x_3-axis. Then the covariant components $\bar{x}_i^k(u, v)$ and $\bar{x}_i(u, v)$ of the vectors $\mathfrak{x}^k(u, v)$ and $\mathfrak{x}(u, v)$ representing H'_k and H' satisfy strictly elliptic Darboux equations (see the preceding), provided $\bar{x}_i$ is of class C^2.

The previously mentioned estimates for the second derivatives cannot be applied to the $\bar{x}_i^k$ because the assumption on the boundary B there must be replaced by the hypothesis that the distance δ_k of H'_k from the boundary of H_k is bounded away from 0. Nevertheless *a priori bounds for the second derivatives also exist under this hypothesis*, see P.[3], pp. 100–106, whose derivation we omit because the calculations are involved. They depend on the same quantities as before with $\inf \delta_k$ replacing the numbers relating to B, and also on the heights of the caps H_k. We emphasize in particular that it is at this place where the *fourth derivatives* of the E_{n_k}, F_{n_k}, G_{n_k}, and hence implicitly those of E, F, G enter.

If we now pass to a subdomain H'' of H' which contains q and whose closure lies in H' then it is easily verified that (23.7) can be applied with $H' = U$ and $H' = D$ to the Darboux equations for the $\bar{x}_i^k$, because they contain only the second derivatives of the E_{n_k}, F_{n_k}, G_{n_k}. The quantities under 3) in (23.7) are uniformly bounded because of the uniformly elliptic character of the equations for the $\bar{x}_i^k$ in H'.

We conclude from (23.7) that the second derivatives of $\bar{x}_i^k(u, v)$ satisfy in H'' uniform Hölder conditions. Therefore the $\bar{x}_i(u, v)$ are of class C^2, and by (23.6) they are of class $C^{m+\alpha}$, $m \geqq 4$, if the intrinsic metric is of class $C^{m+\alpha}$. This finishes the proof of (23.2) and (23.3).

We notice the important corollary:

(23.9) *If the convex surface* K *is intrinsically isometric to a regular convex surface* K' *of class* C^m, $m \geqq 5$, (analytic) *then* K *is regular and of class* C^{m-1} (analytic).

For the intrinsic metric of K' is regular and of class C^{m-1} or analytic.

CHAPTER V

Conclusion

Among the questions suggested by the theories developed in this book two stand out. The first is this:

I.

Although our realization theorems are very general in some respects, they seem special in others. For example, topologically, the only surfaces realizable as complete convex surfaces in E^3 are the sphere, plane, and cylinder. The topological structure of manifolds[1] with indefinite curvature is very general. Can Alexandrov's methods be used to *imbed some of these manifolds as convex surfaces in other three dimensional spaces than* E^3? The concept of convexity is most natural in spaces where the lines have, locally, the properties of projective lines. Therefore we raise this question for three dimensional spherical spaces S_K and hyperbolic spaces H_K with curvature K. The Gauss curvature of a (smooth) convex surface in S_K or H_K is at least K. Therefore the only complete convex surfaces in S_K are, topologically, spheres. But in H_K, where $K < 0$, convex surfaces may have negative curvature and we may expect interesting results. For a precise formulation we need two definitions.

A convex set in H_K is, of course, a set which contains with any two points the hyperbolic segment connecting them. In analogy to our procedure in Section 1, we *define a complete convex surface in* H_K *as the boundary* $B(C)$ *of a three dimensional convex set* C *in* H_K *provided* $B(C)$ *is non-empty and connected.* This requirement eliminated in E^3 for C only the whole space and sets bounded by two parallel planes. In H_K it eliminates many more

[1] We always mean two-dimensional manifolds.

cases. Nevertheless, there is a great variety of convex surfaces in H_K:

The complete convex surfaces in H_K coincide, topologically, with the connected open subsets of the two-sphere.

This follows almost at once from the Klein model of H_K, where the interior $\Sigma_{i=1}^3 x_i^2 < 1$ of $S : \Sigma_{i=1}^3 x_i^2 = 1$ appears as H_K with the open euclidean segments with end points on S as geodesics. If $B(C)$ is a convex surface in H_K then projection of $B(C)$ on S from an interior point of C yields a homeomorphism of $B(C)$ with a connected open subset of S.

Conversely, if such a set G on S is given then we may assume that G is not the whole sphere and also that $S-G$ does not lie in a plane, for we can always find a set homeomorphic to G on S with this property. The convex closure of $S - G$ as set in E^3 has a boundary which, after removal of the points on S, is homeomorphic to G.

Besides the concept of convex surface in H_K we need the definition of a manifold with curvature $\geqq K$. With the angle $\delta(S, T)$, [2] see (18.2), we say in analogy to Section 18:

A manifold with an intrinsic metric has curvature at least K, $K < 0$, if every point has a neighborhood such that for each triangle T in this neighborhood the excess $\epsilon(T)$ is at least as large as the excess of a triangle with the same sides in a hyperbolic plane of curvature K.

Then the following theorem, corresponding to (18.3) and (20.1), holds:

A connected open subset of the two-sphere which is provided with a finitely compact intrinsic metric of curvature $\geqq K$, $K < 0$, can be realized as a complete convex surface in H_K.

There is also a local realization theorem like (19.1). For all these questions we refer the reader to A., pp. 450–462.

II.

The manifolds with curvature $\geqq K$, even if K is allowed to vary, do not comprise all interesting metrics. In particular,

[2] This angle is defined by means of the euclidean Law of Cosines, which is permissible because H_K is, infinitesimally, euclidean.

among the polyhedral metrics, they include only those with non-negative curvature, because the total angle at any point of a convex surface in H_K never exceeds 2π. The second question is, whether the manifolds with non-negative curvature have an extension which includes at least all polyhedral metrics and in which the results of classical differential geometry hold in properly modified forms, as in Sections 14 and 15. It was mentioned in Section 2 that the first generalization of convexity which comes to mind, namely finiteness of the order, proves inadequate.

A convex surface has bounded total curvature, namely, at most 4π. Postulating that the total absolute curvature be locally bounded leads to a fruitful theory.

As pointed out earlier, the upper angle $\delta''(S, T)$ in Section 18 proves preferable in this theory, because the distinction between strong and ordinary angle becomes unnecessary. The excess $\epsilon(T)$ of a geodesic triangle T is defined in terms of this angle. Speaking of total curvature presupposes, at least implicitly, the existence of arbitrarily fine triangulations. Under the assumption of locally bounded curvature (see following), these do exist, but our proof in Section 14 does not apply because it is based on the inclusion property, which no longer holds. For a simple example take a cone with apex a and complete angle 3π, or curvature $-\pi$, at a and on it three segments $T(a, b_i)$, $i = 1, 2, 3$, which form the angle π with each other. Then $T(b_i, a) \cup T(a, b_k)$ is for $i \neq k$ a segment from b_i to b_k. This example shows also that we must be careful in handling geodesic triangles: the triangle $b_1b_2b_3$ has no interior point. Therefore, Alexandrov calls two triangles non-overlapping, if they can be enclosed in sets homeomorphic to closed disks which have no common interior points and then defines:

A manifold M with an intrinsic metric has locally[3] *bounded curvature, if for every compact subset C of M a number $\beta(C)$ exists such that for any finite set of non-overlapping triangles $T_1, \ldots, T_n$ in C*

$$\sum_{i=1}^{n} |\epsilon(T_i)| \leq \beta(C).$$

[3] For brevity A. omits "locally."

Recently and still without published proofs, Zalgaller [1] has reduced the hypotheses in a remarkable way: it suffices to consider only convex triangles homeomorphic to disks and, above all, to restrict only $\Sigma\, \epsilon(T_i)$, which means that the troubles arise only from triangles with positive excess. We call total absolute curvature of a polyhedral metric the sum of the absolute values of the curvatures of its corners; the *positive part of the curvature* is the sum of the curvature of the corners with positive curvature. The two prinicpal theorems of Zalgaller are these:

In a manifold M with an intrinsic metric $\varrho(x, y)$ let every point have a neighborhood U such that for any finite set $T_1, \ldots, T_n$ of non-overlapping convex triangles in U homeomorphic to a disk $\Sigma_{i=1}^{n} \epsilon(T_i) \leqq \beta(U)$, then every compact domain C on M lies in a polygonal domain P so that the metric ϱ induced by ϱ_P in P can be uniformly approximated by polyhedral metrics ϱ_n in P whose total absolute curvatures are uniformly bounded.

The second theorem is a strengthened converse:

Assume that every point of a manifold M with an intrinsic metric ϱ has a neighborhood U with this property: there are polyhedral metrics ϱ_n defined in polygonal domains $P_n \subset U$ which tend, uniformly in U, to ϱ and such that the positive part of the curvature of ϱ_n in U is uniformly bounded for all n. Then M has locally bounded curvature.

That locally bounded curvature is equivalent to the existence of uniformly approximating polyhedral metrics with uniformly bounded total absolute curvatures, had been proved previously by Alexandrov, see A., pp. 491–514 for this and the following statements.

On manifolds, with locally bounded curvature the intrinsic Gauss curvature $\varepsilon(N)$ for Borel sets N, swerves, quasigeodesics, etc., can be defined by appropriate modifications of the procedure used for manifolds with non-negative curvature. There is also a general gluing theorem which guarantees that gluing together pieces of manifolds with locally bounded curvature produce again such a manifold.

Major realization theorems as in the preceding chapter have not been established for the simple reason that no such theorems are known for surfaces with indefinite curvature even under the classical conditions. Among the problems which have been studied we mention the following as examples:

The existence of isothermic parameters, (Reshetnyak [1]). If G is a domain homeomorphic to an open disk (on a manifold with locally bounded curvature) whose boundary has a swerve of bounded variation, then coordinates u, v can be defined in G such that the line element takes the form

$$ds^2 = \lambda(u, v)(du^2 + dv^2),$$

where $\lambda(u, v)$ is the difference of two subharmonic functions.

Surfaces in E^3 whose intrinsic metrics have locally bounded curvature: 1) Surfaces which can, locally, be represented in the form $x_3 = f(x_1, x_2)$, where $f(x_1, x_2)$ is the difference of two convex functions. These surfaces have nearly all the differentiability properties of convex functions, see A. [7]. 2) Surfaces $\mathfrak{x}(u, v)$ of class C^1 with $\mathfrak{x}_u \times \mathfrak{x}_v \neq 0$, where $\mathfrak{x}(u, v)$ has square summable generalized second derivatives in Sobolev's sense. Bakelman [1] studies these surfaces in great detail and derives practically all classical results, even analogues to the Gauss-Codazzi equations. 3) Surfaces with bounded extrinsive curvature, Pogorelov [9].

III.

Some mathematicians will ask why we are interested in these generalizations. The single fact that the manifolds with locally bounded curvature include the polyhedra (besides all convex surfaces) will suffice to make them important and natural to a true geometer. Evidently this argument will not convince those who consider surfaces of class C^∞ as natural geometric objects.

However, there are other reasons for this extension of the concept of surface in differential geometry. Alexandrov and Strel'cov have given a number of examples of *interesting extremum problems which can be formulated, but not solved, within the framework of classical differential geometry*. We will conclude this book with a typical example which illustrates again the efficiency of the method of cutting and gluing. Under the restriction to convex surfaces the example is found in A., pp. 415–417, the general case in Strel'cov [1].

The problem is a *generalization of the isoperimetric problem.* If the length L of a simple closed curve C in E^3 is given then there is no finite upper bound for the areas of surfaces S homeomorphic to a disk with C as boundary. Clearly, a restriction on the curvature of D is necessary. It looks as though we have to restrict the total absolute curvature; surprisingly enough *it suffices to restrict the positive part of the curvature.* Since the area of a right circular cone with a fixed base becomes infinite with its altitude, the curvature must be bounded away from 2π. Thus our problem is:

To find among surfaces homeomorphic to closed disks with boundaries of a given length L and positive part of the curvature at most $\nu^+ < 2\pi$ those which maximize the area A.

We will see that the answer is the right circular cone with curvature ν^+. Hence we have the interesting *isoperimetric inequality*

$$A \leqq L^2\, 2^{-1}\, (2\pi - \nu^+)^{-1}$$

with equality only for a right circular cone of curvature ν^+.

Actually we have not given an exact definition of the positive part of the curvature. But we will limit ourselves to solving the problem for polyhedra whose number of vertices does not exceed a given $n \geqq 4$, for which we did define the positive part of the curvature. We will find that the right pyramid over a regular $(n-1)$-gon and with curvature ν^+ maximizes the area and omit the proof of the plausible fact that a limit process yields the solution for the general case.

We deal with abstract polyhedra of the following type: the boundary C is a simple closed geodesic polygon of length L. The vertices of the metric are, besides the corners in the interior, those points q on C where arcs of C contiguous to q do not form the angle π.

The existence of a maximizing polyhedral metric P with at most n vertices follows from the Gauss-Bonnet theorem which gives $-w^+ - n\pi$ as lower bound for the negative part of the curvature. It is also obvious that the angles at the vertices of P on C measured in P cannot exceed π, because otherwise we could shorten the boundary without increasing ν^+. We remember, see Section 14, that the vertices on the boundary do not contribute to the curvature $\nu(P)$ of P.

We show first that P *cannot have corners with negative curvature.* If such a corner existed we could draw a segment $T = T(b, a)$ from a vertex b of P on C to a. Because the angles of C in P do not exceed π, the segment T does not contain a point of C different from b. We know that T cannot pass through a corner with positive curvature, but it might pass through one with negative curvature. In that case we replace a by the first such corner encountered on T when coming from b. Then we draw a segment $T(a, c)$ so short that it does not contain any corner but a and such that it forms at a equal angles with $T(a, b)$. This common angle α exceeds π because the curvature is negative at a. The case $\alpha \geqq 2\pi$ requires some additional considerations for which we refer, for simplicity, to Strel'cov [1].

We construct a rhombus c_1, a_1, b_1, a'_1 in E^2 such that

$$|a_1 - c_1| = |a'_1 - c_1| = ac, \; |a_1 - b_1| = |a'_1 - b_1| = ab$$
$$\angle\, c_1 a_1 b_1 = \angle\, c_1 a'_1 b_1 = 2\pi - \alpha.$$

We cut P open along $T(b, a) \cup T(a, c)$ and glue the rhombus into the slit such that c_1 falls on c, b_1 on b and $a_1 a'_1$ on the two copies of a. The total angles at a_1 and a'_1 are now 2π and c has become a corner with the negative curvature

$$-\angle\, a_1 c_1 a'_1 = -[2\pi - 2(2\pi - \alpha) - \angle\, a_1 b_1 a'_1]$$
$$= 2\pi - 2\alpha + \angle\, a_1 b_1 a'_1,$$

whereas $2\pi - 2\alpha$ was the curvature at a.

The new polyhedral metric has not more vertices than P (b might have disappeared as vertex), the area has increased, and the positive part of the curvature as well as the length of the boundary are the same as for P. This contradicts the maximizing property of P.

Thus P does not have any corners with negative curvature. Assume there are two corners a_1, a_2 with positive curvatures ν_1, ν_2. A segment $T = T(a, b)$ lies in the interior of P because the angles at the boundary are less than π. We then repeat the construction of Section 17, Figure 10: we form triangles $a'_1 a'_2 a'$, $a''_1 a''_2 a''$ in E^2 with $|a'_1 - a'_2| = |a''_1 - a''_2| = a_1 a_2$ and angles at a'_1, a''_1 equal to $\nu_1/2$ and at a'_2, a''_2 equal to $\nu_2/2$. This is possible because, by hypothesis, $\nu_1 + \nu_2 \leqq \nu^+ < 2\pi$. We cut P open along T and insert the two triangles exactly as in Section 17. Then the points a_1, a_2 are no longer corners, but the point $a' = a''$ is a corner with curvature:

$$2\pi - 2(\pi - \nu_1/2 - \nu_2/2) = \nu_1 + \nu_2,$$

so that ν^+ has not changed, but the area is greater than that of P. This contradicts again the maximizing property of P.

Thus P has at most one corner, and of positive curvature $\leqq \nu^+$. It is now a matter of elementary trigonometry to verify that, for given $\nu^+ > 0$ and sufficiently large n, the maximum is attained when C is a regular $(n - 1)$-gon (of length L) and P a right pyramid with curvature ν^+ at its apex.

LITERATURE

Some of the frequently used references are abbreviated in the text. The abbreviations are given here in bold face.

ALEXANDROV, A. D.

[1] Zur Theorie der gemischten Volumina von konvexen Körpern, Russian, German summaries, parts I, II, III, IV.

I Verallgemeinerung einiger Begriffe der Theorie der konvexen Körper, Mat. Sbornik N.S., **2** (1937), pp. 947–972, **A. [1, I]**

II Neue Ungleichungen zwischen den gemischten Volumina und ihre Anwendungen, *ibid.*, pp. 1205–1238, **A. [1, II]**

III Die Erweiterung zweier Lehrsätze Minkowskis über die konvexen Polyeder auf beliebige konvexe Flächen, Mat. Sbornik N.S., **3** (1938), pp. 27–46, **A. [1, III]**

IV Die gemischten Diskriminanten und die gemischten Volumina, *ibid.*, pp. 227–251, **A. [1, IV]**

[2] Almost everywhere existence of second differentials of convex functions, Russian, Leningrad State Univ., Annals [Uchenye Zapiski] Math. Ser., **6** (1939), pp. 3-35, **A. [2]**

[3] An application of the theorem on the invariance of domain to existence proofs, Russian, English Summary, Izvestiya Akad. Nauk SSSR Sci. Mat., **3** (1939), pp. 243–255, **A. [3]**

[4] Existence and uniqueness of a convex surface with a given integral curvature, Doklady Akad. Nauk SSSR, **35** (1942), pp. 131–134, **A. [4]**

[5] Smoothness of the convex surface of bounded Gaussian curvature, Doklady Akad. Nauk SSSR, **36** (1942), pp. 195–199, **A. [5]**

[6] Additive set functions in abstract spaces, Chapter IV, Mat. Sbornik N.S., **13** (1943), pp. 169–238, **A. [6]**

[7] On surfaces representable by differences of convex functions, Russian, Izvestiya Akad. Nauk Kasah. SSR, **1949**, No. 3, pp. 2–19, **A. [7]**

[8] Convex Polyhedra, Russian, Moscow—Leningrad, 1950, **A. [8]**

[9] A theorem of triangles in metric spaces and some applications, Russian, Trudy Mat. Inst. Strklov, **38** (1951), pp. 5–23, **A. [9]**

[10] Die innere Geometrie der konvexen Flächen, Berlin, 1955, Translation (with supplement) of the Russian book with the analogous title of 1948, **A.**

[11] Uniqueness theorems in the large for surfaces I, Russian, Vestnik Leningrad. Univ., **1956,** No. 19, pp. 1–17.

BAKEL'MAN, I. YA
[1] Differential geometry of smooth non-regular surfaces, Russian, Uspehi Matem. Nauk, **11** (1956), pp. 67–124.

BERNSTEIN, S.
[1] Sur la généralization du problème de Dirichlet, Math. Ann., **69** (1910), pp. 82–136.

BOL, G.
[1] Beweis einer Vermutung von H. Minkowski, Abh. Math. Sem. Univ. Hamburg, **15** (1943), pp. 37–56.

BONNESEN, T. and FENCHEL, W.
[1] Theorie der konvexen Körper, Berlin, 1934, or New York, 1948, **K.**

BOULIGAND, G.
[1] Introduction à la géométrie infinitésimale directe, Paris, 1932.

BUSEMANN, H.
[1] Intrinsic area, Am. Math., **48** (1947), pp. 234–267.
[2] The isoperimetric problem for Minkowski area, Amer. J. Math., **71** (1949), pp. 743–762.
[3] The Geometry of Geodesics, New York, 1955.

BUSEMANN, H. and FELLER, W.
[1] Krümmungseigenschaften konvexer Flächen, Acta Math., **66** (1935), pp. 1–47, **B. F. [1].**
[2] Bemerkungen zur Differentialgeometrie der konvexen Flächen, parts I, II, III.
I Kürzeste Linien auf differenzierbaren Flächen, Mat. Tidsskr., **B1935**, pp. 25–36, **B. F. [2, I].**
II Über die Krümmungsindikatrizen *ibid.*, pp. 87–115, **B. F. [2, II].**
III Über die Gaussche Krümmung, Mat. Tidsskr., **B 1936**, pp. 41–70, **B. F. [2, III].**
[3] Regularity properties of a certain class of surfaces, Bull. Amer. Math. Soc., **51** (1945), pp. 583–598, **B. F. [3].**

BUSEMANN, H. and MAYER, W.
[1] On the foundations of calculus of variations, Trans. Amer. Math. Soc., **49** (1941), pp. 173–198.

COHN-VOSSEN, S.
[1] Zwei Sätze über Starrheit der Eiflächen, Nachr. Ges. Wiss. Göttingen Fachgruppe, **I** (1927), pp. 125–139.

COURANT, R. and HILBERT, D.
[1] Methods of Mathematical Physics, vol. I, New York, 1953.

DARBOUX, G.
[1] Théorie générale des surfaces, vol. III, Paris 1894, vol. IV, Paris, 1896.

DEHN, M.
[1] Über die Starrheit konvexer Polyeder, Math. Ann., **77** (1916), pp. 466–473.

DENJOY, A.
[1] Sur l'intégration des coefficients différentiels d'ordre supérieur, Fund. Math., **25** (1935), pp. 273–326.

FENCHEL, W.
[1] Inégalités quadratiques entre les volumes mixtes des corps convexes, C.R. Acad. Sci. Paris, **203** (1936), pp. 647–650.
[2] Généralizations du théorème de Brunn et Minkowski concernant les corps convexes, *ibid.*, pp. 764–766.

FENCHEL, W. and JESSEN, B.
[1] Mengenfunktionen und konvexe Körper, Danske Vid. Selsk. Mat.-Fys. Medd., **16, 3** (1938), pp. 1–31.

GROTEMEYER, K. P.
[1] Gleitverbiegungen und eindeutige Bestimmtheit isometrischer ebenrandiger Mützen, Math. Z., **59** (1953), pp. 278–289.

HAHN, H. and ROSENTHAL, A.
[1] Set Functions, Albuquerque N.M., 1948.

HARTMAN, P.
[1] On local uniqueness of geodesics, Amer. J. Math., **72** (1950), pp. 723–730.

HARTMAN, P. and WINTNER, A.
[1] On the problems of geodesics in the small, Amer. J. Math., **73** (1951), pp. 132–148.

HAUPT, O.
[1] Über die Krümmung ebener Bögen endlicher Ordnung, S.B. phys-mediz. Soc. Erlangen, **17** (1939), pp. 219–227.

HERGLOTZ, G.
[1] Über die Starrheit der Eiflächen, Abh. Math. Sem. Univ. Hamburg, **14** (1942), pp. 127–129.

HILBERT, D.
[1] Grundzüge einer allgemeinen Theorie der linearen Integralgleichungen, Leipzig—Berlin, 1912.

HJELMSLEV, J.
[1] Grundlag for Fladernes Geometri, Mem. Acad. Roy. Sc. Let. Danmark. Copenhague, (**7**) XII (1914).

HOPF, E.
[1] Über den funktionalen, insbesondre den analytischen Charakter der Lösungen elliptischer Differentialgleichungen zweiter Ordnung, Math. Z., **34** II (1931), pp. 194–233.

HOPF, H. and VOSS, K.
[1] Ein Satz aus der Flächentheorie im Grossen, Arch. Math., **3** (1952), pp. 187–192.

JESSEN, B.
[1] Om konvekse Kurver's Krumning, Mat. Tidsskr., **B 1929**, pp. 50–62.

LEIBIN, A.
[1] On the deformability of convex surfaces with a boundary, Russian, Uspehi Matem. Nauk N.S., **5** (1950), pp. 149–159.

LEWY, H.
[1] On differential geometry in the large I (Minkowski's problem), Trans. Amer. Mat. Soc., **43** (1938) pp. 258–270.
[2] On the existence of a closed surface realizing a given Riemannian metric, Proc. Nat. Acad. Sc. U.S.A., **24** (1938), pp. 104–106.

LIBERMAN, J.
[1] Geodesic lines on convex surfaces, Doklady Akad. Nauk SSSR, **32** (1941), pp. 310–313.

LYUSTERNIK, L. and SCHNIRELMAN, L.
[1] Topological methods in variational problems and their application to the differential geometry of surfaces, Russian, Uspehi. Matem. Nauk N.S., **2** (1947), pp. 166–217.

MARCHAUD, A.

[1] Sur les continus d'ordre borné, Acta Math., **55** (1930), pp. 67–115.

MINAGAWA, T. and T. RADÓ

[1] On the infinitesimal rigidity of surfaces, Osaka Math. J., **4** (1952), pp. 241–286.

MINKOWSKI, H.

[1] Volumen und Oberfläche, Math. Ann., **57** (1903), pp. 447—495; Ges. Abhandlungen, vol. II, Leipzig—Berlin, 1911, pp. 230–276.

NIRENBERG, L.

[1] On non-linear elliptic differential equations and Hölder continuity, Communications in Pure and Appl. Math., **6** (1953), pp. 103–156.

[2] The Weyl and Minkowski Problems in differential geometry in the Large, *ibid.*, pp. 337–394.

OLOVYANISHNIKOV, S. P.

[1] On the deformation of unbounded convex surfaces, Russian, Mat. Sbornik N.S., **18** (1946), pp. 434–440.

[2] Generalization of a theorem of Cauchy on convex surfaces, Russian, *ibid.*, pp. 441–446.

POGORELOV, A. V.

[1] Unique determination of convex surfaces, Russian, Trudy Mat. Inst. Steklov, **29**, 1949, 98 pp., **P. [1]**.

[2] Quasigeodesics on convex surfaces, Russian, Mat. Sbornik N.S., **25** (1949), pp. 275–306.

[3] Deformation of Convex Surfaces, Russian, Moscow, 1951, **P. [3]**.

[4] The regularity of a convex surface with a given Gauss curvature, Russian, Mat. Sbornik N.S., **31** (1952), pp. 88–103.

[5] On the rigidity of convex polyhedra, Russian, Zapiski Mat. Ot. Fiz-Mat. Fak. H G U, **23** (1952), pp. 79–89, **P. [5]**.

[6] On the question of the existence of a convex surface with a given sum of the principal radii of curvature, Russian, Uspehi Matem. Nauk, **8** (1953), pp. 127–130.

[7] Die eindeutige Bestimmtheit allgemeiner konvexer Flächen, Schrift. Forsch. Inst. Math., **No. 3, 1956.** Translation of the Russian book with the analogous title of 1952, **P. [7]**.

[8] On the inflexibility of general convex surfaces with curvature 2π, Russian, Doklady Akad. Nauk SSSR, **106** (1956), pp. 19–20.

[9] Surfaces of bounded extrinsic curvature, Russian, Izdat. Har'kovskogo Ordena Trud. Kras. Znam. Gosud. Univ. Im. Gor'kogo, 1956.

RADON, J.
[1] Theorie und Anwendungen der absolut additiven Mengenfunktionen, S.B. Akad. Wiss. Wien Abt, **IIa 122** (1913), pp. 1295–1438.

REIDEMEISTER, K.
[1] Über die singulären Randpunkte eines konvexen Körpers, Math. Ann., **83** (1921), pp. 116–118.

RESHETNYAK, YU. G.
[1] Isothermic coordinates in manifolds with bounded curvature, Russian, Doklady Akad. Nauk, **64** (1954), pp. 631–634.

[2] On a generalization of convex surfaces, Russian, Mat. Sb. N.S., **40** (82) (1956), pp. 381–398.

SIEGEL, C. L.
[1] Integralfreie Variationsrechnung, Nachr. Akad. Wiss. Göttingen Math.-Phys. Kl. **IIa** (1957), pp. 81–86.

STEINITZ, E. and RADEMACHER, H.
[1] Vorlesungen über die Theorie der Polyeder, Berlin, 1934.

STREL'COV, V. V.
[1] Estimates of the length of a polygon on a polyhedron, Russian, Izvestiya Akad. Nauk Kasah. SSR, **1952,** No. 116; Ser. Astr. Fiz. Mat. Meh., **1** (6), pp. 3–36.

VOSS, K.
[1] Einige differentialgeometrische Kongruenzsätze für geschlossene Flächen und Hyperflächen, Math. Ann., **131** (1956), pp. 180–218.

WEYL, H.
[1] Über die Bestimmung einer geschlossenen konvexen Fläche durch ihr Linienelement, Vierteljschr. Naturforsch. Ges. Zürich, **61** (1916), pp. 40–72.

ZALGALLER, V. A.
[1] On the foundations of the theory of two-dimensional manifolds with bounded curvature, Russian, Doklady Akad. Nauk, **108** (1956), pp. 575–576.

INDEX